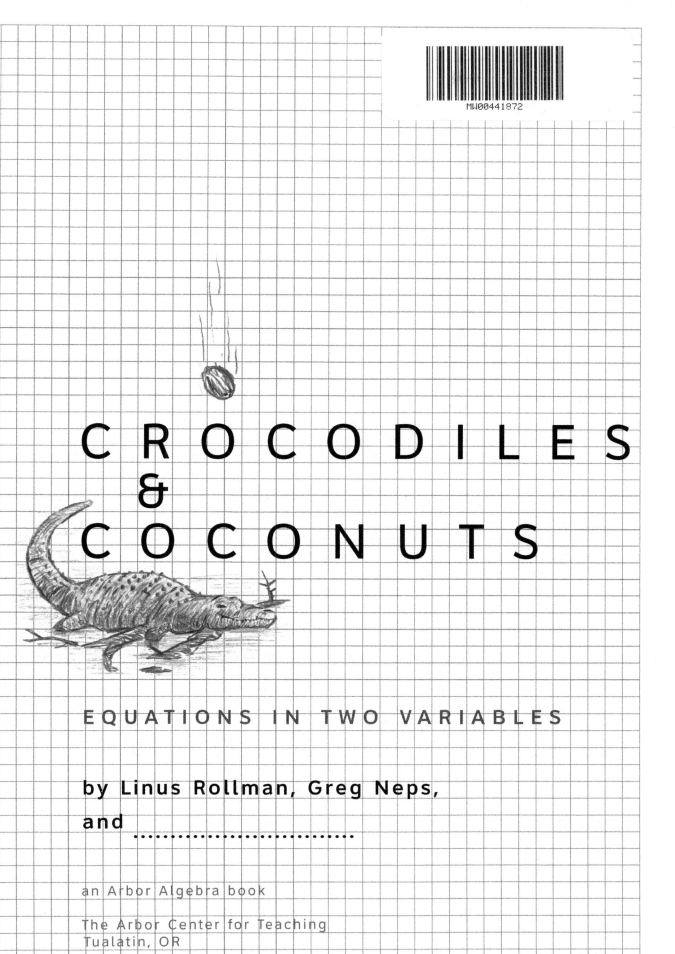

CROCODILES
& COCONUTS

EQUATIONS IN TWO VARIABLES

by Linus Rollman, Greg Neps,
and ....

an Arbor Algebra book

The Arbor Center for Teaching
Tualatin, OR

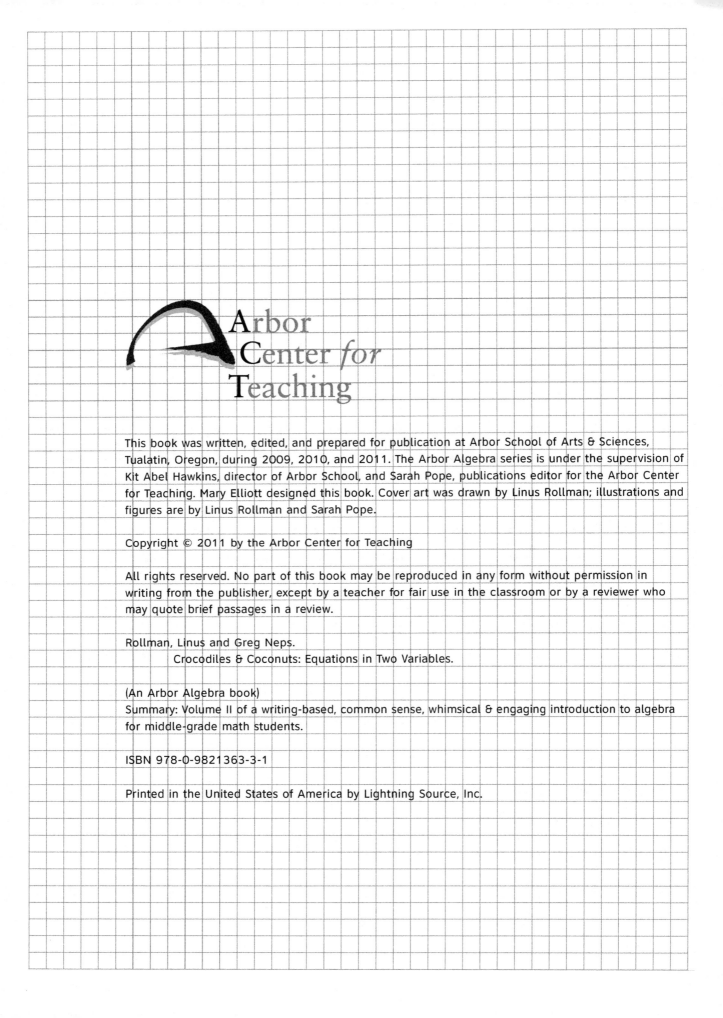

Arbor Center for Teaching

This book was written, edited, and prepared for publication at Arbor School of Arts & Sciences, Tualatin, Oregon, during 2009, 2010, and 2011. The Arbor Algebra series is under the supervision of Kit Abel Hawkins, director of Arbor School, and Sarah Pope, publications editor for the Arbor Center for Teaching. Mary Elliott designed this book. Cover art was drawn by Linus Rollman; illustrations and figures are by Linus Rollman and Sarah Pope.

Rollman, Linus and Greg Neps.
     Crocodiles & Coconuts: Equations in Two Variables.

(An Arbor Algebra book)
Summary: Volume II of a writing-based, common sense, whimsical & engaging introduction to algebra for middle-grade math students.

ISBN 978-0-9821363-3-1

Printed in the United States of America by Lightning Source, Inc.

TABLE OF CONTENTS

The Arbor Class of 2012,
whose effort, persistence, and humor
were invaluable in crafting this book:

ALY
DANIEL
LILY
JONATHAN
HILLARY
GRACE
ANNA
CAROLINE
LIA
CHARLIE
NOAH
HANNAH
SARA
NAT
MAX
GUTHRIE
MCKELLAR
COURTNEY
ISABELLA
SIERRA

O
THE CROW
&
THE PITCHER

An aged crow was flying through a parched landscape, and he himself was parched — indeed, he was dying of thirst. Suddenly his sharp old eyes spotted something strange: below him was a table, and on the table was a pitcher. And in the pitcher there was water! He flew down and landed on the table, delighted and relieved. But soon his delight turned to despair: although there was enough water in the pitcher to save him, the water level was low and the neck of the pitcher was too narrow for him to get his head inside. Even with his long beak, he was unable to so much as taste the water. This was the end. He was going to perish within a few centimeters of his salvation. If only he had been an old armadillo, with a long snout! He very nearly gave up and collapsed on the spot. Thankfully, though, he had a brave heart hidden in his chest, and moreover he was very clever. He sat and he thought and he thought and he sat, always staring at the pitcher. At last, unable to bear the sight of that water, he turned his back on it. And that was when he noticed something: the ground around the table was covered with tiny little pebbles, each almost perfectly round. An idea came into his clever old head, and it was his salvation. He began to flit back and forth between the table and the ground. With each flit, he carried a pebble in his beak, and each pebble he dropped into the pitcher. Slowly, slowly, the level of the water in the pitcher began to rise. At last (he had lost count of how many pebbles he used) the water rose high enough that he could reach his beak down and drink. What a glorious drink it was! The old crow was saved.

The fable I've just told you is based on one originally written by Aesop, and now I'm going to ask you to use that fable as the basis for an experiment that has to do with algebra.

Here's what you will need to collect in order to perform the experiment:

- A clear cylinder that can stand up on its own and that is the same width all the way up. (In other words, it can't be vase shaped, or even pitcher shaped.) A tall glass might work pretty well, but it's nice if you have something you can write on. An empty tennis ball tube works particularly well.

- A marker to write on the side of your cylinder. If your cylinder is glass, you'll want a piece of masking tape to put on the side. Then you can write on the masking tape.

- A ruler with centimeter markings

- A bunch of marbles of uniform size

- Some graph paper and a pencil for recording your results and answering the questions that go with this experiment

The first step is to make centimeter marks along the side of your cylinder, going up (so you can measure the height of the water that's in it). You actually don't need to worry about starting exactly at the bottom of the cylinder (which is a little tricky with a tennis ball tube) since you're going to be measuring the change in the water's height.

Now fill your cylinder with water up to an even centimeter mark. Leave enough room in it so that the water level can rise as you drop marbles in.

The next thing that you'll do is drop in marbles and measure the change in the water level. But before you start, I'd like you to think about how to record your data. I would strongly suggest using a table that looks something like this (only probably bigger):

Number of marbles (x)	0									
Height of water in cm (y)										

Notice that, in the table I've just shown you, the letters **x** and **y** are included in parentheses. This being an algebra book, those are variables. You don't need to worry about them just this moment, but be prepared for the fact that once you're done with the experiment, I'll start talking about using **x** to stand for the number of marbles and **y** to stand for the height of the water.

There are two ways that you can go about conducting this experiment. You can either drop in the marbles one at a time and record the height of the water as you go, or you can drop in however many marbles are necessary to raise the height of the water by one centimeter and then make a recording. I'd recommend the second method, even though you may have to fudge a little bit, since a whole number of marbles may not raise the level exactly a centimeter.

Go for it. Stop before the water spills.

All right, now I have some questions for you.

1. Remember that I told you that we'd be using x to stand for the number of marbles and y to stand for the height of the water? When we're dealing with two-variable situations (which is what we'll be doing a lot in this book), one variable is called the *dependent variable* and the other is called the *independent variable*. Thinking about the experiment that you just did and the meaning of the word "dependent," do you think that x or y is the dependent variable in this case? Why?

2. Now I'm going to ask you to make a graph of the data that you collected. I'm sure that you've done this before, in science classes, for instance. The horizontal edge of your graph is, in this case (and this is the standard algebraic way of doing things), going to be called the *x-axis*. On this axis, you'll keep track of the x-variable: how many marbles you put in. What do you suppose the vertical edge of your graph will be called and what will you keep track of on it?

3. Go ahead and set up your graph. So that you can easily and accurately compare your own graph with those of your classmates, use standard units: 1 unit of graph paper should stand for 1 centimeter on the y-axis and 1 unit of graph paper should stand for 2 marbles on the x-axis. Carefully put the points from your table onto your graph.

4. Now connect each point on the graph to the next point. How would you describe the shape of the graph?

5. Where does the line that you made touch the y-axis? How many marbles had you put in at the point where the line touches the y-axis?

6. Written as a fraction, what is the ratio of the height increase of the water to the number of marbles you added? Another way of asking this question is to say, "How many marbles are necessary to raise the water level by one centimeter?" and then write that as a fraction with the centimeter as the numerator and the number of marbles as the denominator.

The ratio that you just found is very important. You can call it the *rate of change* (in centimeters per marble). If, for example, the ratio you got was **1/3**, this tells you very quickly that it takes three marbles to raise the water level by one centimeter. Looking at it the other way, one marble raises the water by one third of a centimeter.

At this point, you'll need to get together with your classmates and compare your results. You'll need to do Problems 7 through 12 with at least a couple of other people. Your teacher will help you decide how big those working groups should be.

7. By examining someone else's graph, how can you determine the water level that he or she started with?

8. By examining someone else's data table, how can you determine the water level that he or she started with?

9. How many marbles did it take to raise each of your partners' water levels by one centimeter?

10. How many marbles would it take to raise each group member's (including your own) water level by 2.5 cm? How did you calculate this?

11. Compare all of the graphs in your group. It's probably the case that the lines on these graphs slope upward at different rates — that some are steeper than others. Discuss with one another why you think that this might be the case. It will probably help to describe to each other how you did the experiment. Record the ideas that you come up with.

12. Based on the work that you did, name at least one thing that might affect how steep a graph is. In what ways do those things affect the graph of an experiment?

13. If you dropped 1,000 marbles into your "pitcher" (assuming it was tall enough), estimate how high the water would get. How did you come up with this estimate?

14. What if you dropped in a million marbles? How did you make that estimate?

There is a way to use the work that you've done on this experiment and your algebra skills to make an exact prediction for the answers that you just estimated in Problems 13 and 14. It involves writing a *formula* — and it's the reason I asked you to use the variables **x** and **y**. It's also at the very heart of the work that you'll be doing throughout most of this textbook. A formula is an equation that, in this case, relates the variables **x** and **y**. In order to be most useful, your formula should have the form **y = _____**.

15. Your job is to write the formula by filling in the blank in the statement above. In order to do this, look at the table of data you collected, remember that y stands for the height of the water and x stands for the number of marbles, and ask yourself the question, "For each pair of numbers in the table, what do I have to do to x to get y?" What you will find is that you have to do the exact same thing to each x-value in order to get the corresponding y-value. In fact, in each case, you'll be multiplying x by a number (remember what happens when you multiply by a number that's less than one!) and adding another number. Go ahead and complete the formula.

Often, when you are dealing with a formula or an equation of the kind that you just wrote, it's handy to think of it in this way:

y = (the rate of change)x + the initial starting point

Or, in English, "**y** equals the rate of change times **x** plus the initial starting point."

16. Compare your formula to the answer you found for Problem 6. What's the relationship between the rate of change and the ratio you found in Problem 6?

You can now use the formula you've written to predict the height of water in your "pitcher" for any number of marbles. Simply replace the **x** in your formula with the desired number of marbles and do the calculation to find **y**.

17. Use this method to find the height of the water for 1,000 marbles and for 1,000,000 marbles. How close are these values to the estimates that you made in Problems 13 and 14?

The wonderful thing about all this is that algebra (your formula) can be used to describe something in the real world (the way the height of the water in your "pitcher" changed as you added marbles) and can be used to make predictions about things that can happen in the real world. This fact is incredibly useful. It has helped make possible all sorts of things that we take for granted, from building bridges to flying airplanes to performing heart surgery. But before we get too carried away...

18. Is there anything unrealistic about the situations that you used your formula to predict in Problem 17? What might actually happen if we were to try to push this particular experiment to those extremes?

19. What do you think the upper limits might be on the number of marbles and the height of the "pitcher" where the formula you found still applies?

1

TWO-VARIABLE EQUATIONS & THE CARTESIAN COORDINATE PLANE

1 WHAT DOES IT MEAN TO SOLVE A TWO-VARIABLE EQUATION?

In the last textbook, you got very good at manipulating equations with one variable. When I say "manipulating," what I mean is that you found equivalent equations in order to discover the value of the variable. The big shift in this textbook is that you will be working with equations that contain two variables. I know that this may not sound like a very big deal — in fact it changes everything, as I hope you may have begun to see as you worked on the experiment in the Introduction. Before you dive into two-variable equations, why don't you start by solving the following one-variable equations just to warm up? Write down all of the steps that you use!

1. $3x + 6 = 15$

2. $21 = 5x - 4$

3. $2x + 3 = 3x - 5$

4. $3(x + 4) = 13$ (I recommend using the Distributive Property as a first step for this one.)

5. $-2(x + 3) = 4(x - 6)$

6. $4(-x + 6) = -3(x + 8)$

7. $13 - 3(x - 2) = -4(-x - 3)$

8. $5.2x - 1.6 = 6.4x + 2$

Excellent. Remember when you first started working with single-variable equations and I asked you to picture them as pan balances? Boxes stood for variables and circles stood for numbers. For example, the pan balance picture that would go with Problem 1 would look like this:

You learned to solve single-variable equations by taking imaginary weights off of those pan balances or shifting imaginary weights around. Just to remind you how that worked, try it with Problem 1.

9. **Go back to the steps that you used to solve Problem 1 and make pan balance drawings to go with each step.**

The basic rule for solving single-variable equations was this: *any time you do something to one side of the balance, you have to do an identical thing to the other side*. The same rule continues to apply with equations that have two variables (or, for that matter, three, four, five and so on… but we'll only worry about two).

Let's look at a two-variable equation. Say, **2y = 4x + 6**. I'm going to draw a pan balance picture for this equation. Just as in the last book, little circles will stand for numbers and boxes will stand for **x's** — but this time I'll add triangles that will stand for **y's**.

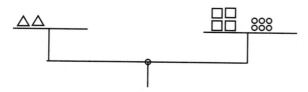

All this means is that you have **2** of one kind of mystery weight that balance with **4** of another kind of mystery weight plus **6** one-pound weights. As long as I do the same thing to both sides, I don't change the balance — in other words, I get an *equivalent equation*. For example, I could add **4** one-pound weights to each side, in which case the picture would look like this:

… and the equation that would go with it would be **2y + 4 = 4x + 10**.

I could also, say, double both sides. But I'll leave that one to you.

10. **What equivalent equation do you get when you double both sides of 2y = 4x + 6? If it helps you to draw a pan-balance picture, go ahead and do so.**

11. **What equivalent equation do you get when you subtract 10 from each side of 2y = 4x + 6?**

12. **Find eight more equations that are all equivalent to 2y = 4x + 6. Make sure that you use all four operations (multiplication, division, addition, and subtraction) while you're creating those eight equations.**

Great. One of the most important algebraic skills is simply being able to play around with equations, changing them from one form to another. And really, in this regard, two-variable equations are just like one-variable equations. So let's start talking about the differences.

13. **What does it mean to "solve" a single-variable equation? (Hint: There should be a page in your Note to Self book that has nothing but the answer to this question on it.)**

14. Based on what it means to "solve" a single-variable equation, go ahead and try to "solve" the equation 2y = 4x + 6. (Spend at least five minutes working on this. If you really try to solve it and think about it carefully, it will tell you something very important about two-variable equations.)

Personally I think the concept that you just began to investigate is kind of crazy and wonderful, and understanding it is key to taking the next steps in algebra.

Remember that solving a single-variable equation means finding the value or values of the variable that make the equation true. Most single-variable equations that you worked with in the last book had just one solution. There are three important things to understand right away about two-variable equations. The first has to do with *how many* solutions they have.

15. Find as many values for x and y as you can that make the equation 2y = 4x + 6 true. (Don't spend more than five minutes on this. Remember that you can use negative values for x and y!)

I'm not sure how many values for **x** and **y** you found that made the equation **2y = 4x + 6** true. Maybe you found ten, maybe you found a hundred. But I can guarantee that you did not find all of the values that make that equation true. You could spend the rest of your life working on that problem and never find them all, because there are *infinite* sets of values for **x** and **y** that make it true. Not only can you go on forever making **x** and **y** larger and larger or smaller and smaller, but, in case you didn't try this...

16. Check to see whether the values y = 6 and x = $\frac{3}{2}$ make the equation 2y = 4x + 6 true.

In other words, since **x** and **y** can be fractions, not only can **x** and **y** both keep getting more and more positive or both keep getting more and more negative, they can both be split into pieces as well. As I said, there are infinite pairs of values for **x** and **y** that will make the equation **2y = 4x + 6** true. In fact, all two-variable equations have infinite sets of solutions.

The second thing to understand about two-variable equations is related to the first.

17. Check to see whether the values y = 5 and x = 2 make the equation 2y = 4x + 6 true.

18. Find five pairs of values for x and y that DO NOT make the equation 2y = 4x + 6 true.

So, even though there are infinite pairs of **x** and **y** values that make the equation **2y = 4x + 6** true, not just *any* pair of **x** and **y** values will make it true. Indeed, there are infinite pairs of **x** and **y** values that *do not* make the equation true.

(We don't need to go really deeply into this right now, but this brings up a couple of fascinating points. How is it possible for a set of numbers to be infinite but not include all numbers? Isn't anything that's infinite, well, infinitely big? Also, if you picked a random pair of values for **x** and **y**, chances are that they would not make the equation **2y = 4x + 6** true. So it seems probable that there are more pairs that *don't* make it true than pairs that *do* make it true. There are also, however, infinite sets of each. Does that mean that one infinity can be bigger than another infinity?)

The third important thing to know about two-variable equations is something that you have undoubtedly noticed: their solutions come in pairs of numbers. You need to have a value for **x** and a value for **y** to make the equation **2y = 4x + 6** true. The way that mathematicians traditionally write solutions to two-variable equations is like this: **(1, 5)** The first number in the parentheses is the value for **x** and the second number is the value for **y** that goes with it. The pair **x = 1** and **y = 5** makes the statement **2y = 4x + 6** true and is therefore one of its infinite pairs of solutions.

19. In this lesson, you learned three things about two-variable equations that are important enough to merit a *Note to Self*. It could be called something like *Characteristics of Two-Variable Equations*.

The next two problems will ask you to solve a puzzle and write a couple of equations, each with two variables. The point is partly to play around with some algebraic thinking, partly to see that two-variable equations can involve things other than pitchers and marbles. Also, we'll come back to how to use algebra to solve this kind of puzzle later in the book.

20. I have 29 coins (all nickels and dimes) totaling $2.15. How many nickels do I have? I'm going to ask you to guess and check in order to solve this problem and to track your guessing and checking in a chart. I'd recommend one that looks something like the one below. (I've put one example set in — though it's a lousy guess!)

Number of nickels	Number of dimes	Number of coins (total)	Value (in cents)
100	2	102	520

21. Now the more important part. Look back at the original problem and use it to write two different equations relating the number of nickels to the number of dimes. Use x to represent the number of nickels and y to represent the number of dimes. I predict that one of these equations will be pretty easy for you to write and one (that has to do with the value of the coins) will be harder. Get help if you need to and check to make sure that your two equations really work. We'll come back to this later on in this book.

REVIEW

Simplify the following expressions. Show all of the steps that you use. For example, the steps for simplifying the expression 3^2 might look like this: $3^2 = (3)(3) = 9$

1. $(5)^2$

2. $(3x)^2$

3. $(-8)^2$

4. $-(8)^2$

5. -8^2

6. Problems 3, 4, and 5 all look similar, but one has a different answer from the others... why?

7. The first angle of a triangle is five times as large as the second. The third angle is half the sum of the other two. What is the measure of each angle? It will probably help you to draw a picture, to think in terms of ratios, and to use a variable and write equations. An important hint: The sum of the three angles of a triangle is 180°.

8. I had $960 to spend on a computer and accessories. I spent two-thirds of it on the computer and $120 on a printer. How much money do I have left?

Perform the following operations:

9. $(3.2)(2.7)$

10. $0.057 + 9.86$

11. $\dfrac{56.42}{0.07}$

2 THE COORDINATE PLANE

1. The first thing that I'm going to ask you to do in this lesson is to play a game. It's called Crocodiles & Coconuts and it bears a close resemblance to a game you may have played before, only with a couple of important differences. You'll need a piece of graph paper, a pencil, and a partner.

First, you'll set up two gameboards on your sheet of paper — one for recording the positions of your own crocodiles and one for recording your guesses at the locations of your opponent's crocodiles. The gameboard consists of two crossed number lines, one horizontal and one vertical. They should each be numbered from -8 to 8 (the horizontal one from left to right and the vertical from down to up) and they should cross each other at the zero mark. In other words, they ought to look about like this:

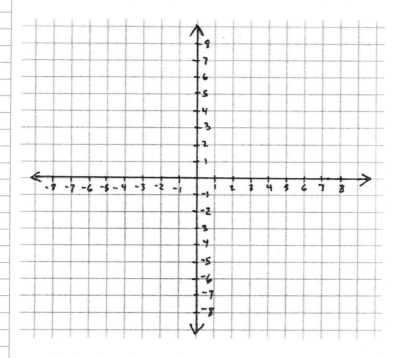

Remember, you should have two gameboards on your sheet. Now, on one of those gameboards, choose and record the positions of your crocodiles — one big croc five units long, two crocs four units long, and two crocs three units long. The key here is that the crocodiles occupy intersections on the gameboard. In other words, one possible placement of a five-unit croc might look like this:

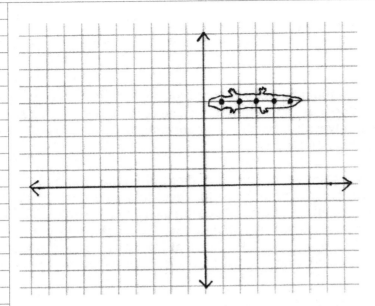

This crocodile requires five hits to knock it out — one coconut to the snout, two to the tail, and two to its middle. Your crocs can be placed vertically or diagonally as well as horizontally.

Now you can start. One player makes a guess that represents dropping a coconut at that intersection and the other player tells him or her whether it was a hit or a miss. Then it's the second player's turn to drop a coconut. When you make a guess, say the number on the horizontal number line first and then say the number on the vertical number line. If your opponent said, "Five, negative three," you'd find that intersection by starting at the spot where the number lines cross, moving five units over to the right and then three units down. The winner is the player who knocks out all of his or her opponent's crocodiles first.

Play at least two games.

It is probably not obvious what the game of Crocodiles & Coconuts has to do with what you were doing in the last lesson. I hope that the connection will become clearer in the next chapter, because it's very important. (Keep it in mind over the course of this chapter — maybe you can figure it out. Think about the Crow and the Pitcher and what you know about graphing.)

The Crocodiles & Coconuts gameboard is really something that is known in mathematics as the *Cartesian coordinate plane* (or often just "the coordinate plane"). It's named after a Frenchman named René Descartes (pronounced ruh-NAY day-CART; *Cartesian* is car-TEE-zhan) who lived about four hundred years ago and who is credited with inventing it. Actually, as is the case with very many mathematical and scientific concepts, it's not necessarily one hundred percent clear who was responsible for coming up with the concept of the coordinate plane. The fact is that people tend to share ideas and build on each other's ideas and, usually, more than one person deserves credit for coming up with something. Nevertheless, there's no doubt that Descartes was brilliant. (He did a lot of work, by the way, outside of the field of mathematics; he was particularly interested in trying to prove the existence of God.) There's also no doubt that the coordinate plane is a fantastically useful mathematical concept.

One thing to understand about the coordinate plane is that no matter the size of the drawing, you're always supposed to imagine that the two number lines stretch on into infinity in all four directions. In order to indicate this, the plane is drawn with arrows at the ends of the number lines, just as in the illustrations that went with Crocodiles & Coconuts.

Now there is some vocabulary to take care of. First off, the point where the two number lines cross is known as the *origin*. (I think that this is actually kind of a poetic name — poetic naming not necessarily being something mathematicians are good at.) The horizontal number line is called the **x-axis** and the vertical number line is called the **y-axis**.

The power of the coordinate plane lies in the fact that you can use two numbers to specify any point on a flat surface, just as you did when you said "five, negative three" while you were playing Crocodiles & Coconuts. These two numbers are called the *coordinates* of that point. The number on the **x-axis** (called the **x-coordinate**) is said or written first and the number on the **y-axis** (called the **y-coordinate**) is said or written second. Traditionally, a set of coordinates is written like this: **(5, -3)**

The four sections of the coordinate plane are called *quadrants*. The upper right quadrant is called the *first quadrant* (often referred to with a Roman numeral: I) and the numbering proceeds counter-clockwise, like so:

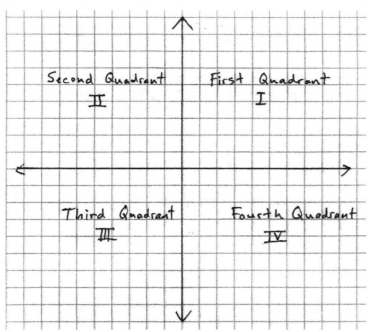

The order in which the quadrants are numbered is one of those things that you just have to memorize.

By the way, you don't need to number every point along the two axes. One or two points on each branch usually suffice. Also, it's perfectly okay for a point to be *on* either the **x-** or **y-axis**, and therefore to be on the border between two quadrants. Here is a graph of the point **(0, 5)**:

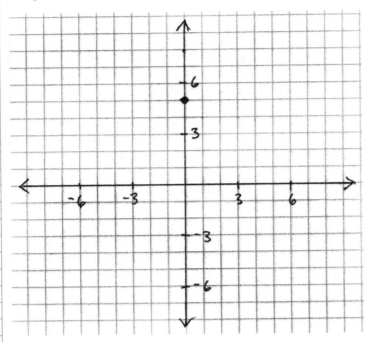

One last fact about the coordinate plane and then I'll ask you to practice using it. Most of the time, in most math books (including this one), the points that we use will have whole number coordinates, like **(5, -3)** or **(7, 15)** or **(-9, -4)**. This is really just for convenience, because points with whole number coordinates are easy to plot. However, there is no reason that points can't have fractional or decimal coordinates. You just have to make estimates for the fractional components. For example, the points **A (2.5, -3.5)** and **B (-5 ³/₄, 4 ¹/₂)** could be plotted like this:

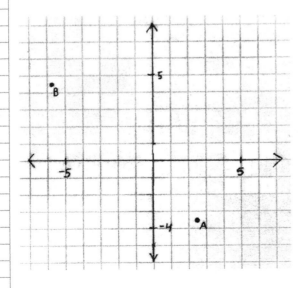

2. Here's a little Cartesian coordinate practice devised by my student Hannah Park. You'll need to make a grid that extends to positive and negative 15 on each axis. Mark each of the following coordinate points. As you mark each point, connect it to the last one with a straight line. (Before you start, here's a reminder: Each pair consists of an x-coordinate followed by a y-coordinate, so that's *first* left or right, *then* up or down.)

(-6, 4) (0, -13) (6, 4) (10, 15) (6, 6) (6, 14) (5, 7) (5, 11) (4, 5) (3, 15) (2, 9) (2, 12) (1, 5) (1, 14) (0, 6) (0, 10) (-1, 5) (-2, 13) (-3, 6) (-3, 11) (-4, 8) (-5, 15) (-5, 9) (-6, 11) (-5, 5) (-10, 14) (-6, 4) (6, 4)

Start a new line: (-2, 2) (-1, 1) (-2, 0) (-3, 1) (-2, 2)
Start a new line: (2, 2) (1, 1) (2, 0) (3, 1) (2, 2)
Start a new line: (0, -2) (1, -1) (-1, -1) (0, -2)
Start a new line: (-2, -3) (-1, -5) (0, -4) (1, -5) (2, -3) (1, -4) (0, -3) (-1, -4) (-2, -3)

3. Create your own picture puzzle using the Cartesian coordinate plane like the one that you did in Problem 2. Give it to a friend or family member to solve.

4. Write a **Note to Self** explaining the *characteristics of the coordinate plane*. Make sure to include the vocabulary that you learned in this lesson, particularly *origin*, *quadrant*, and *axis*.

5. Salt Lake City is laid out on a coordinate graph with the Temple at the origin. So you get addresses like 330 S, 400 W for a building. Every algebra student should be put there for a morning to find addresses. There's a map of Salt Lake City on the next page. Find and mark the addresses as accurately as you can. Find the axes first. Notice that the addresses don't necessarily adhere to the mathematical convention of listing the x-coordinate first and don't let that mess you up.

Salt Lake City Addresses:

a.	Allstate Insurance	261 E, 300 S
b.	The Diamond Source	230 W, 200 S
c.	Bride's Shop	430 East South Temple
d.	Chameleon Artwear	1065 E, 900 S
e.	Grunts and Postures	779 E, 300 S
f.	House of Guitars	645 S, 300 W
g.	Mechanized Music	511 W, 200 S
h.	Acoustic Music	857 E, 400 S
i.	Liza's Pizza	716 E, 400 S
j.	Fendall Ice Cream Co.	470 S, 700 E
k.	Sharper Image	602 E, 500 S
l.	Spencer's for Steaks and Chops	255 South West Temple
m.	Train Shoppe	470 S, 900 E
n.	Golden Braid Books	151 S, 500 E
o.	East High School	840 S, 1300 E
p.	West High School	241 N, 300 W
q.	Albertsons	140 N, 900 W
r.	Smith's Food and Drugs	455 S, 500 E
s.	Maverick Country Store	615 S, 200 W
t.	Salt Lake Regional Medical Center	1050 East South Temple

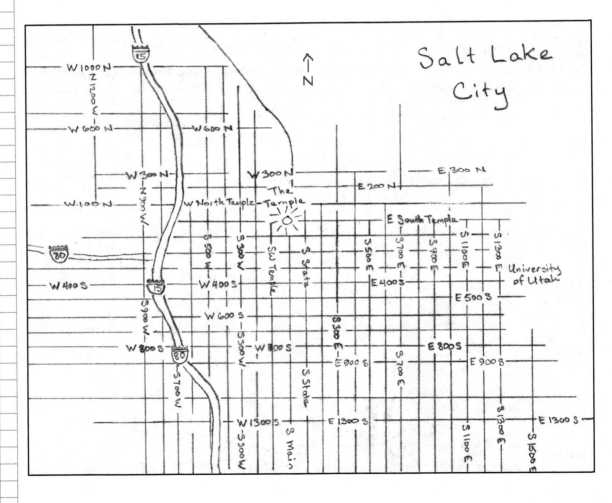

REVIEW

Simplify the following expressions. Look back at your Note to Self on the Order of Operations from *Jousting Armadillos* if you are having difficulty. (Remembering the acronym PEMDAS may help.)

1. $3 + -2$
2. $-8 - (-6)$
3. $-18 - 4$
4. $-9 + 13$
5. $-17 - (-4)$
6. $32 + (-9)$
7. $-14 + (-6)$
8. $(8^2 - 4^2)$
9. $(8 - 4)^2$
10. $3 - 2^2$
11. $6 - 2(5 + 4)$

12. $[12 + (4 - 3)2] + \dfrac{12}{3}$

13. $\dfrac{[5(8 - 3) + (27)(3)]}{(3^4 - 5^2)}$

14. $|-3(2) - 4(5)| - 6(-2)$

15. Find the positive-integer common multiples of 8 and 12 smaller than the product of 8 and 12.

16. The length of a rectangle is equal to twice its width. If the perimeter is 42 meters, what is its length? You might try drawing an illustration, but you can also solve it algebraically. Try writing an equation that relates l and w.

17. A survey found that about 45 out of 500 people polled are left handed. If that holds true for the general population of a school that has 200 people, including teachers, how many people at the school would you expect to be sinistral? What percent is that? What do you think might be the relationship between left-handedness and the word *sinister*?

18. You arrive after dark at the Mythological Zoo. Luckily, there are still some pretty cool exhibits to see at night. The sign outside one of the exhibits reads, "Cerberus and Polyphemus: Three-Headed Dogs and One-Eyed Men. Occupancy: 32 Creatures." As you shine your flashlight into their habitat you catch the reflection of 137 glowing eyes. How many of each creature does the exhibit contain?

3 LINES IN THE COORDINATE PLANE

In this lesson and the next, you'll be looking at some of the characteristics of lines in the coordinate plane. After that, we'll connect this coordinate plane business back to the algebra that you've been learning.

1. You'll probably need an entire blank page in your notebook to do this problem. Make a coordinate plane that shows -14 to 14 on each axis. Mark two points on the plane as follows: the points should not be on the same vertical or horizontal line and the right-hand point should be higher up than the left-hand point. Label the left-hand point A and the right-hand point B. Record the coordinates of the two points somewhere to the side.

2. Imagine that you're giving someone directions to get from point A to point B. Suppose that the person has to stay on the whole number grid lines and that you want to give the simplest possible directions. What would they be?

3. I think there were really only two options for the last question. Either you told the person to go right a certain number of spaces and then up a certain number of spaces, or you told them to go up a certain number of spaces and then right a certain number of spaces. Now I'd like you to use a ruler or some other straightedge to connect the two dots with a line (directly — a diagonal line) and extend that line to roughly the edges of your coordinate plane; we'll call this line AB. Now, over to the side where you recorded the coordinates of A and B, record the directions that you gave to the person as a ratio, with the vertical movement as the numerator and the horizontal movement as the denominator. You can call this ratio m (for "movement," if you like). Simplify this fraction as much as possible. To clarify, here's an example of what your page might look like:

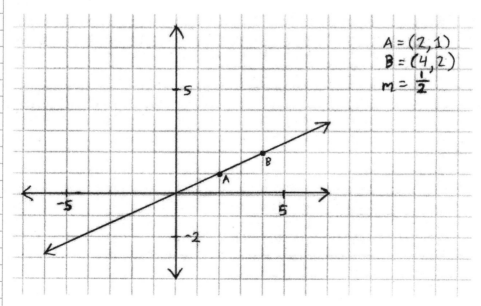

4. Repeat the entire process for four more sets of points. (They can be C and D, E and F, and so on.) This will give you four more lines on your page. Make sure that the right-hand point is always higher than the left-hand point and try to do a bunch of different kinds of lines — that is, make them different levels of slantiness and have them be in different quadrants and cross between quadrants. Record the coordinates and the m-values for each line next to the graph. Work neatly because otherwise your page may get a little confusing.

5. Now compare all of your lines and their m-values. What do you notice about the m-value of a line and its slantiness?

The technical word for "slantiness" is *slope*. I'm sure you've run across that word before, for example when someone was describing a hill — this is the same idea. The **m-value** is the slope of a line, and the higher the **m-value**, the steeper the line. It makes sense — if the ratio of how far you move up to how far you move over is very high, that means you've climbed a long way up and only gone a little way over.

6. Examine the following graph. Use points L and M to find the slope exactly as you did for the lines that you drew.

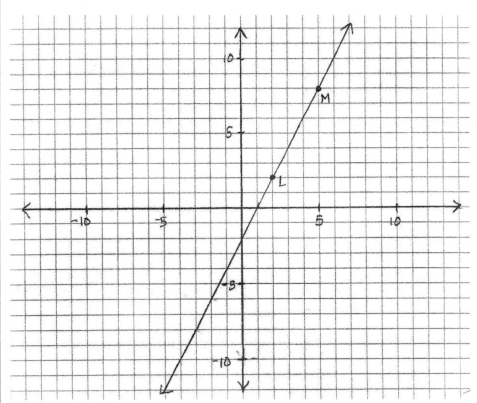

7. Now choose any two other points on that line. (The easiest spots will be where it crosses the intersections of the graph-paper lines.) Use those two points to find the slope. Choose two more points and find the slope. Choose one last set of two points and find the slope.

8. What can you conclude about the slope of a straight line?

9. Now look back at the coordinates of the points that you wrote down for your lines in Problem 4. It's possible to determine the slope of a line just from the coordinates of the two points. How can you do it? (This is not as hard as it may sound. The question is really just this: How can you look at the coordinates of the two points and judge how far up and how far over you'd have to move in order to go between them? It will be easiest for points that are both in the first quadrant, since these don't involve any negative numbers.)

Your explanation probably went something like, "Take the **y-coordinate** of the one point and subtract the **y-coordinate** of the other point, then divide that by the **x-coordinate** of the first point minus the **x-coordinate** of the second point." That's a fine explanation, but it could be a little confusing, so I'm going to translate it into a mathematical formula, a sort of recipe for finding the slope. (You wrote a formula for the height of the column of water when you worked on the Crow and the Pitcher. There will be more about formulas later on.) In order to use this formula, you just need to know that the symbol x_1 means the **x-coordinate** of the first point and the symbol x_2 means the **x-coordinate** of the second point. Same goes for the symbols y_1 and y_2. So here's the formula:

$$m = \frac{y_2 - y_1}{x_2 - x_1}$$

You may ask yourself, "So which one is the first point and which is the second point?" The answer is that it doesn't matter, as long as you're consistent.

10. In order to verify the statement that I just made, use the formula to calculate the slope of good old line AB. Do it twice — once with point A as the "first point" and once with point A as the "second point." (Just be careful about subtracting and dividing with negatives. If you need a reminder of how to do it, check your Note to Self book from *Jousting Armadillos*.)

By the way, you may sometimes hear or read that "slope is equal to rise over run." This is just another way to phrase the formula you just learned. *Rise* means vertical change ($y_2 - y_1$) and *run* means horizontal change ($x_2 - x_1$). So rise over run is just:

$$\frac{y_2 - y_1}{x_2 - x_1}$$

11. On a fresh coordinate plane, create a line that you are sure — without calculating — will have a slope of 2.

12. On the same plane, create lines that you're sure will have slopes of 3 and of 1/3.

13. On the same plane, create a line that has a slope of 1 and passes through the origin (the point (0, 0) where both horizontal and vertical number lines begin).

Use the formula that you learned to determine the slope of the lines in the problems below. Show all your work.

14. The line that passes through (3, 2) and (5, 3)

15. The line that passes through (2, 4) and (3, 5)

16. The line that passes through (-1, 3) and (2, 4)

17. A line that hits the y-axis at -4 and passes through the point (5, 3)

18. The line that passes through (-3, -2) and (1, 4)

19. Write a *Note to Self* that explains what the *slope of a line* is and how to find it using the formula.

REVIEW

Simplify or solve the following:

1. $\dfrac{36}{54}$

2. $\dfrac{75}{125}$

3. $\dfrac{64}{128}$

4. $\dfrac{35x}{105}$

5. $\dfrac{12x}{52}$

6. $\dfrac{12}{48}x$

7. $\left(\dfrac{2}{3}\right)\left(\dfrac{3}{7}\right)$

8. $\left(3\tfrac{1}{2}\right)\left(2\tfrac{3}{4}\right)$

9. $\dfrac{3}{5} + \dfrac{5}{9}$

10. $5\tfrac{1}{4} - 7\tfrac{1}{2}$

11. What is one-tenth of one-tenth?

12. $(0.2)(0.2)$

13. You and a group of weasels are sharing a cake. You get 30% of the cake. But you want to share your part of the cake with your other buddies, a group of aardvarks. After you finish sharing it, you're left with only 20% of the portion that the weasels gave you. What percent do you have of the original cake?

14. One liter of juice concentrate is mixed with four liters of water.
a. What is the ratio of juice concentrate to water?
b. What fraction of the mixture is juice concentrate?
c. What percent of the mixture is juice concentrate?

15. A cab driver charges $8 for each cab ride, plus $3.25 per mile. Find the total cost of a 9-mile trip.

16. If a person earns $32.50 in five hours, how much will he earn in 8 hours?

17. A rectangle's length is 7 more than 3 times its width. If the perimeter is 78 meters, what's the rectangle's length?

18. Linus bought his new motorcycle in Washington and had to pay 6% sales tax on the cost of the bike. The total cost was $6,360. What was the price of the motorcycle before sales tax?

MORE ABOUT SLOPE

1. **Look at the following line:**

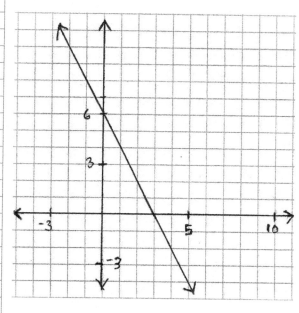

Choose any two points on the line and, using the formula you learned in the last lesson, find its slope.

2. Do the same thing for each of the following lines.

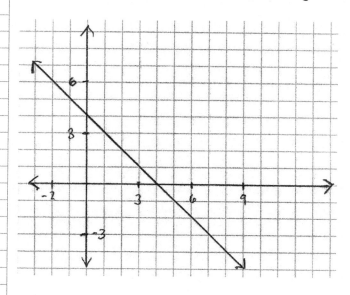

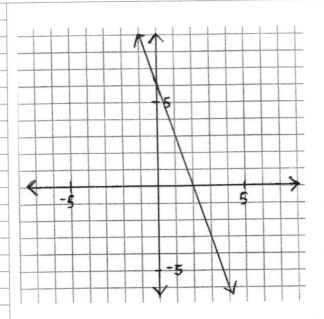

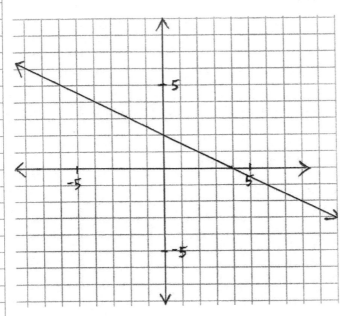

3. Compare the four lines that you just analyzed to the lines that you drew in the last lesson. Although each line is different from the others, the lines from this lesson all have a common characteristic. The lines in the last lesson also had a common characteristic. What is it that makes the lines in this lesson different from the lines in Lesson 3?

Now compare the values of the slopes from this lesson to the values of the slopes from the last lesson. What can you conclude (using inductive reasoning!) about the relationship between the sign of a line's slope and the direction in which that line slopes?

Use the same formula you've been using to predict whether the slope of each of the following lines is negative (sloping down from left to right) or positive (sloping up from left to right).

4. The line that passes through (-5, -4) and (-2, -1)
5. The line that passes through (2, 3) and (5, 2)
6. A line that includes the point (3, 1) and intersects the x-axis at 4
7. A line intersecting the y-axis at 5 and the x-axis at -2
8. The line that passes through (-6, -3) and (-2, 8)

9. Take a look at the following lines:

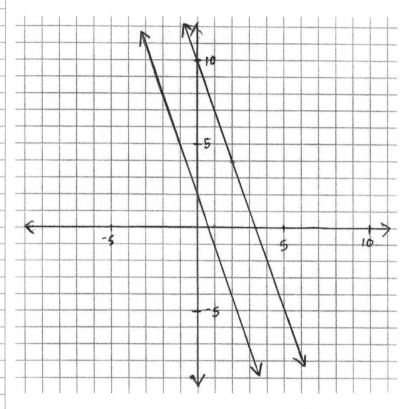

Based on their graphs, you know that the slopes of these two lines will be negative. What else can you predict about their slopes? Test your hypothesis by calculating their slopes.

10. The lines that you worked with in Problem 9 are parallel. What do you think will be the case about the slopes of any parallel lines? Test your hypothesis by graphing at least three pairs of parallel lines and calculating their slopes.

Chapter 1, Lesson 4

11. Take a look at these lines:

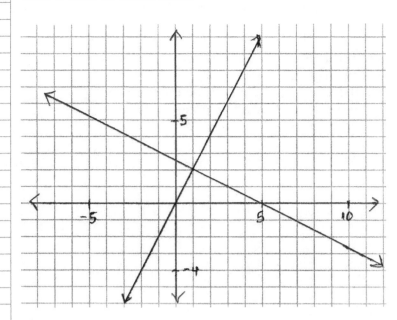

These lines are perpendicular to one another, meaning that where they cross, they form four 90° angles. Think carefully and see if you can predict what the relationships of their slopes will be. (This one is a good deal trickier than the parallel line case from Problems 9 and 10. Don't jump to an immediate conclusion.) Check your hypothesis by calculating the slopes of the two lines. The relationship between their slopes may not be obvious even after you've calculated them, so...

For the next three problems, plot line AB and determine its slope. Then plot line CD and determine its slope. Compare the slope of line AB with that of line CD.

12. A(0, 3) B(4, 6)
C(6, 2) D(3, 6)

13. A(-2, 7) B(3, 9)
C(4, -1) D(9, -3)

14. A line passing through the origin and the point (5, 3)
C(1, 6) D(4, 1)

15. Based on Problems 11 through 14, what can you conclude about the slopes of perpendicular lines? (It may help you to remember that when you were learning to divide fractions, you learned that the upside-down version of a fraction was called its *reciprocal*.)

16. Graph the following pair of lines and explain how their slopes follow the rule you just wrote in Problem 15.

A(5, 0) B(3, -2)
C(-4, 7) D(-2, 5)

17. Look at this line:

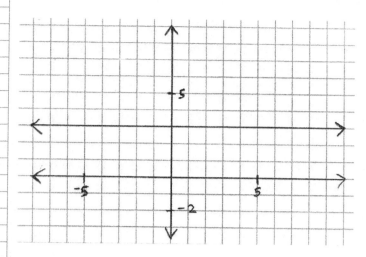

This is, of course, a horizontal line. Predict what its slope will be. Test your prediction, using the formula. Create two more horizontal lines and calculate their slopes. What is the slope of any horizontal line?

18. Now have a look at this one:

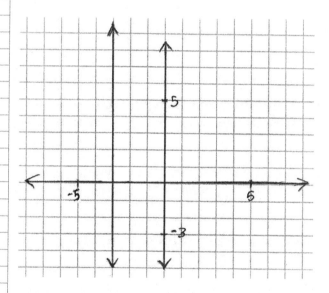

This one's a vertical line; just as you've been doing, predict its slope and test your prediction. Create two more vertical lines and make sure that their slopes are the same as the first one's.

Okay, that last problem was a test for you. If you happily concluded that the slope of a vertical line was zero, you definitely need to look back in your Note to Self book at the section about one and zero. When you try to find the slope of a vertical line, you end up with a zero in the denominator of your fraction, which means that you're dividing by zero, and as you very well know, YOU CANNOT EVER DIVIDE BY ZERO, NO WAY, NOT EVER, PERIOD. In other words, a vertical line has no slope. To put it as a mathematician would, the slope of a vertical line is *undefined*.

Now that you have a pretty solid understanding of how to get from one point to another, let's practice a bit more.

Graph the lines that include the following pairs of points. (You should practice graphing often.) Then determine their slopes.

19. A(3, 1) B(4, 3)
20. A(-2, 1) B(5, 4)
21. A(3, 0) B(4, 5)
22. A(2, 1) B(4, -3)
23. A(5, 2) B(5, -4)

Now practice using the formula to find slopes — no graphs, just algebra.

24. A(3, 2) B(5, 3)
25. A(-1, -2) B(3, 3)
26. A(-3, 1) B(4, -2)
27. A(4, 2) B(4, -1)
28. A(-3, 7) B(2, 7)

29. Look back at the graph you created for Problem 19. Can you create a line parallel to it? What would be the coordinates of two points on a line parallel to AB? What would be its slope?

30. Look back at the graph you created for Problem 23. Can you create a line parallel to it? What would be the coordinates of two points on a line parallel to AB? What would be its slope?

31. Look back at Problem 28. What would be the coordinates of two points on a line parallel to AB? What would be its slope?

32. Bear with me, I know it seems a bit repetitive. Just three more of these. What are the slopes of lines perpendicular to lines AB in Problems 19, 22, and 28?

33. Write a *Note to Self* about *varieties of slope* that explains the difference between positive and negative slope and also tells what you know about the slopes of parallel and perpendicular lines. Be sure to use examples and illustrations.

REVIEW

1. The number of hours left in the day is one-third of the number of hours that have already passed. How many hours are left in the day?

2. Linus is thinking of two numbers. Their greatest common factor is 6. Their least common multiple is 36. One of the numbers is 12. What is the other number? Does he just sit around all day thinking of numbers?

3. Complete the magic square below using the numbers 1-9 exactly once. All rows, columns, and diagonals must have the same sum.

   Hint: Although there are several possible solutions, only one number will work in the middle box.

4. Now it's time to do this in terms of algebra. Draw another magic square like the one above. Set your middle box equal to x and let all the rest represent their relationship to x. What sum do you get when you add any column, row, or diagonal of the square you just completed?

5. Linus and Greg have the same amount of money. If Linus gives 1/3 of his money to Greg, what will be the ratio of Linus's money to Greg's money? Drawing a picture might be an excellent way of figuring out this problem.

6. The local taxidermy store is going out of business. At first they marked everything in the store down 10%. The next week they marked their remaining inventory down an additional 40%. What would be the final sale price of a stuffed moose originally priced at $1,200? Does this strike you as a great deal or a rip-off?

7. Now this is a tough one, but stick with it! You and your friend want to take your speedboat out on the Columbia River for a two-mile trip. You bet your friend that you could average 60 mph. For the first mile you average 45 mph.

   a. How much time do you have left to travel that second mile?
   b. How fast would you have to go to win your bet?

8. What percent of 150 is 12? (Set up a proportion and use a variable to solve this problem.)

2

FUNCTIONS

1 THE PARTS OF A FUNCTION

Having completed the first chapter, you can work comfortably in the Cartesian coordinate plane and know something about the characteristics of straight lines in that plane. Now it's time to figure out what that has to do with algebra.

First, I'd like to remind you that equations with two variables, unlike most single-variable equations, have infinite numbers of solutions, but that not all values for **x** and **y** will make a given equation true.

1. **Look back at your work for Chapter 1, Lesson 1, Problem 15. In that problem, you found as many values as you could for x and y that made the equation 2y = 4x + 6 true. Copy those values into a table like this:**

x	2				
y	7				

(If you don't have at least five sets of numbers on your table, find some more values for x and y that will make the equation true. You don't have to put more than five sets in. If possible, choose pairs where neither x nor y is >12 or <-12.)

2. **Make a coordinate plane that shows -12 to 12 on each axis. Now put the points from the chart you made for Problem 1 on the coordinate plane, exactly as you did in the last two lessons. What do you notice about the points?**

Maybe you had already made this connection between a two-variable equation and the coordinate plane (your work on the Crow and the Pitcher probably revealed it), but it is a very big deal. It is the connection that Descartes made between the coordinate plane and algebra: you can represent the infinite solutions to any two-variable equation on the coordinate plane.

3. **Go back to the graph you made in Problem 2, join all of the points together, and make an arrow at each end of the line, just as you did with the axes of the plane. (Have I mentioned that the plural form of "axis" is "axes," pronounced AX-eez?)**

Your graph now properly represents of all of the solutions to the equation **2y = 4x + 6**. Every single point on that line is a solution to the equation, even the points that lie between graph paper squares. The arrows indicate that the line theoretically goes on forever in both directions, and every point on that infinite line is another solution to the equation.

The three things you have used so far in this lesson — the equation **2y = 4x + 6**, the table of solutions that you made in Problem 1, and the graph that you finished in Problem 3 — are, in a sense, all representations of the same thing. In mathematical terms, that "thing" is called a *function*.

The actual, technical definition of a function is "a pairing of two sets of numbers so that each number in the first set corresponds to exactly one number in the second set." This means that the set of numbers in this table:

x	5	6	52	17	-13
y	7	10	-5	-4	10

... is a function, because for every **x-value** there's only one **y-value**. But the set of numbers in this table:

x	-2	3	3	5	-6
y	12	4	9	16	8

... is *not* a function, because the **x-value 3** corresponds to two different **y-values**. (By the way, don't try to study those two tables for a pattern — as far as I know, there isn't one.)

Don't worry too much about the technical part of the definition. None of the sets of numbers you'll be dealing with will have that problem of a repeated **x-value**, so they'll all be functions. And every really interesting function (like the ones you'll be dealing with) can be represented in three ways: in a table, on a graph, or with an equation (also sometimes called a formula).

The important skill for you to practice for right now is moving between these three representations. The easiest thing to do is to move from a table to a graph or a graph to a table.

When you move from table to graph, as you've already done, you just plot the points onto a coordinate plane. For every table in this book, you can assume that it represents just a few pairs out of an infinite set of possible pairs, so you can go ahead and connect the points with a line and put arrows at the end of the line indicating that it extends forever. (In fact, many two-variable equations don't form straight lines, and your job when you graph one of those is to represent it with a smooth curve. But we won't work on any equations like that just yet.)

Graph the lines that correspond to the tables in Problems 4 through 8. Normally, you'd graph each line on a separate graph, but in this case, graph all five lines on the same graph.

4.

x	0	1	2	3	4
y	0	1	2	3	4

5.

x	0	1	2	3	4
y	1	2	3	4	5

6.

x	0	1	2	3	4
y	-2	-1	0	1	2

7.

x	0	1	2	3	4
y	3	4	5	6	7

8.

x	0	1	2	3	4
y	-4	-3	-2	-1	0

9. In order to answer this question you need to know that the point where a line hits the y-axis is traditionally called the line's *y-intercept*. If I were to ask you to fill in the blanks in the following sentence, could you?

All the lines in Problems 4-8 have the same _____; the only difference between them is their _____.

Moving from a graph to a table is just as simple as moving from a table to a graph. All you have to do is pick several points that lie on the line and record them in a table. (How many is "several?" Well, ideally enough so that if someone put them back on a coordinate plane, that person could tell where the line was supposed to be. One point is obviously too few. Two points are technically enough, as long as the "straight line" part is assumed. Four points is probably about the right number.)

Create tables to go with the graphs in Problems 10 through 12.

10.

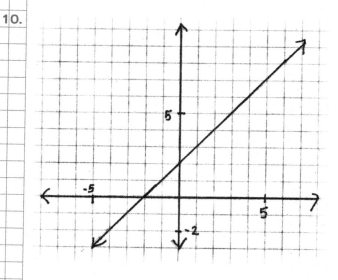

11.

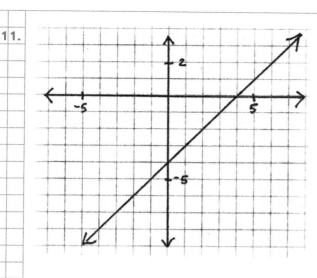

12.

In the next few lessons, we'll work on moving between the other representations of functions.

REVIEW

Show all work.

1. The total cost of a car is $11,247.12, including the optional rocket boosters and ejection seat that you will, of course, want. If the dealer wants you to pay in 36 equal payments, how much will each payment be?

2. A recipe uses two eggs, three cups of flour, and some other stuff. The recipe makes enough batter for six servings. How many cups of flour will you need to make 10 servings?

3. Matt is a great bargain shopper; he found a chipmunk carrier that cost him only $42. It was on a shelf labeled "30% off the lowest marked price," and the lowest marked price was already a 50% reduction from the original price. What was the original price of the chipmunk carrier?

4. Elyse wants to average 90% on her quizzes, and her marks so far are 45/50, 30/40, 23/25, and 34/40. The next quiz has 50 questions. What's the minimum number she needs to answer correctly to get an average of 90%?

5. Greg is a sea monkey farmer. Every year he sells 4.5% of his herd (if that's what you'd call a group of sea monkeys) at the international sea monkey auction. Last year he wrangled 3,500,000 sea monkeys, but a hatch of new baby sea monkeys has increased his herd by 8%. How many sea monkeys can he take to the auction this year?

6. Angela takes home $6,000 per month from her job as a doughnut tycoon. She spends 15% on rent and $100 on utilities. Of the remainder, she spends 10% on health insurance. (You can't be too careful when you eat doughnuts every day.) Then she spends 2/9 of what's left on food other than doughnuts. One seventh of the resulting remainder goes to fuel and expenses for her Porsche. (A tycoon needs to travel in style.) She spends 1/5 of the new remainder on disco outfits and 1/2 of what's left on entertainment (mostly at discotheques). How much can she save each month toward a takeover of the strudel industry?

7. Achilles and the Tortoise are appearing on Dancing with the Stars. They both start at the same point and dance their way in opposite directions around the 106-meter circumference of a circle. The Tortoise can Viennese waltz at 8 meters per second; Achilles can rumba at 10 meters per second, but he misses his cue and starts 2 seconds after the Tortoise. How far will each one have traveled before they run into each other?

2 SLOPE-INTERCEPT FORM

Before I ask you to focus on translating between equations and graphs, I'd like you to know that two-variable equations are often written in a form that looks like this:

y = _____

Equations do not have to be in that form. For example, in the first lessons in this book you worked quite a bit with the equation **2y = 4x + 6**. However, the "**y = _____**" form is extremely useful, for reasons that you may already have figured out and that you'll explore further in this lesson.

Many two-variable equations look something like this:

y = 2x + 7

As you'll see (and may already have realized), the graphs of those equations are straight lines and are therefore called *linear equations*. The generic form of these equations looks like this:

y = mx + b

In this form, **y** and **x** are real variables and **m** and **b** stand for numbers. The **m** is called the *coefficient* and the **b** is called the *constant*.

1. **Write an equation in the form y = mx + b, where the coefficient is six and the constant is ten.**

2. **Write an equation in the form y = mx + b, where the coefficient is negative three and the constant is five.**

3. **Write an equation in the form y = mx + b, where the coefficient is two and the constant is negative nine.**

4. **Write an equation in the form y = mx + b, where the coefficient is one and the constant is zero.**

As you saw in Problem 4, it is possible for **b** to be equal to zero. Mathematicians would count the equation **y = 4x** as being in the form **y = mx + b**; it just has a **b-value** of zero. It's also possible for the **m-value** to be zero. If you multiply anything by zero you get zero, so that means that a mathematician would say that the equation **y = 6** was also in the form **y = mx + b**, only with a value of zero for **m**. I hope you also realized when doing Problem 4 that it is not necessary to write down a coefficient of one. In other words, **y = 1x + 3** and **y = x + 3** mean exactly the same thing.

5. **Explain why y = 1x + 3 and y = x + 3 mean the same thing.**

The **y = mx + b** form that you've been working with is sometimes called the *y-equals form* (for reasons that I expect are obvious) and sometimes the *slope-intercept form* (for reasons that you will soon investigate).

It's important and useful to be able to transform equations into slope-intercept form. Converting an equation to this form is also known as *solving for* **y**. (If you converted an equation to be, say, **x = 5y - 2**, that would be known as *solving for* **x**.) As you work on the following problems, remember the cardinal rule for manipulating equations: whatever you do to one side of the equation (besides simplifying), you must do the same thing to the other side.

Solve the following equations for y.

6. 2x + y = 8   7.   y - 4 = 3x
8. 5 + y = 4x   9.   y - 7 = 2x + 6
10. 3x = -y + 5   11. 6x + y + 2 = 5x - 3

12. 7 + 9x + y = 7   13. $8 + \frac{1}{2}x + y = 0$

14. 3x + 4 - 2x = -y + 8 - 4x   15. 5x - 2 = 1y + 10

There's one complication that you may run across when converting equations to slope-intercept form. Let's consider the equation **12x = 4y + 16**. I'm going to take you through one possible route to solving that equation:

12x = 4y + 16
12x - 4y = 16 (subtract **4y** from each side)
-4y = -12x + 16 (subtract **12x** from each side)

At this point, the next step is to divide each side by **-4**. The one thing you need to remember is that you need to use the Distributive Rule and divide *all* of the parts of the right-hand side by **-4**. The final equation, once you've divided by **-4**, will look like this:

y = 3x - 4

16. **Explain why dividing the equation -4y = -12x + 16 by -4 gives y = 3x - 4. There are three division steps involved — be sure to explain them all.**

When you're working with these kinds of equations, you may get fractional values for **m** or **b**. There is no problem with that — just remember to simplify your fractions as much as possible.

Solve the following equations for y.

17. 2y = 4x + 8   18. 3y + 6x = 12
19. -4y - 20 = 8x + 4   20. 15x + 10 = 5y - 20
21. 3y = 1x + 9   22. 2x + 3y = 2x + 9
23. 7x + 7y = 14   24. 3x + 2y = 10

25. 8x + 9 = 3y - 3   26. $\frac{y}{4} + 3 = 2x + 5$

Chapter 2, Lesson 2

Now that you've had some practice at solving for **y**, it's time to look at why the slope-intercept form is useful in the first place. To do that, I'll ask you to do some graphing based on equations. For these next problems, I'll ask you to first convert the equations into tables. In other words, all you do is find a few pairs of **x-** and **y-values** that are solutions to the equation, then you plot those pairs on a graph and use them to make a line. All of the equations that you're going to work with here are linear equations — their graphs are straight lines — so...

27. How many solutions (points on the line) do you need to find before you can make a graph?

28. Set up your page like this, so that it has a nice, big graph and a chart like the one I've shown underneath it:

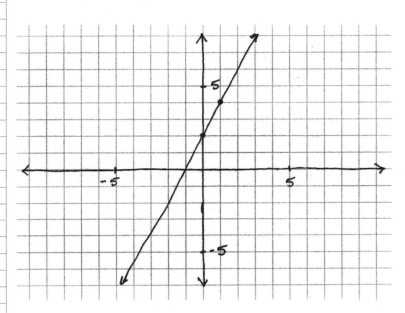

equation	table			slope	y-intercept
y = 2x + 2	x	0	1		
	y	2	4		
y = x + 2					
y = -3x + 2					
y = -2x + 2					

I've begun the work by creating a table for y = 2x + 2 and putting it on the graph. You need to create tables and graphs for each of the other equations — put all the lines on the same graph. Then calculate the slope of each line, either by using the formula or by counting rise over run on the graph, and put those in the chart. Then add the y-intercepts to the chart.

29. What do the four lines from Problem 28 have in common? What makes those four lines different from each other?

30. In the form y = mx + b, how does m affect the graph of a line?

31. Make a graph and chart just as you did for Problem 28, but use these equations:

y = 2x + 1
y = 2x + 3
y = 2x - 2
y = 2x - 4

32. What do the four lines you graphed in Problem 31 have in common? What makes them different from each other?

33. In the form y = mx + b, how does b affect the graph of the equation?

34. Why is the form y = mx + b known as "slope-intercept" form?

35-44. Use what you've just discovered to graph the equations that you simplified in Problems 17 through 26 *without creating tables*. Plot each line on a separate graph.

As you've just seen, slope-intercept form is very handy when it comes to graphing equations. The roles of **m** and **b** in the form **y = mx + b** make a lot of sense, too. Think about **b** — b is the **y-intercept**, right?

45. What is the value of y in the equation y = 2x + 5 when x is zero? What's the value of y in the equation y = 4x - 3 when x is zero? What's the value of y in the equation y = 4x + 1 when x is zero? What's the value of y in the equation y = mx + b when x is zero?

Now think about the role of **m**, which is only slightly more complicated. You've discovered that **m** represents the slope of the line.

46. Consider the equation y = 2x + 4. Make a table showing the values of y in that equation when x is equal to 1, 2, 3, and 4. For the equation y = 2x + 4, how much does y go up every time x goes up by one?

Now consider the equation y = -3x + 1. Make a table showing the values of y in that equation when x is equal to 1, 2, 3, and 4. For the equation y = -3x + 1, how much does y go "up" every time x goes up by one?

Based on what you just saw, in the generic equation y = mx + b, how much does y go up every time x goes up by one?

If **y** goes up by **m** every time that **x** goes up by **1**, that's the same thing as saying that every time the graph goes one step to the right, the graph goes **m** steps up. Well, that's essentially another way of defining the slope of a line. For instance, a line with a slope of **2** goes two steps up for every one step that it goes to the right.

Of course, based on what you now know, it's also easy to go from a graph to an equation. All you have to do is remember that **m** is the slope of the line and **b** is its **y-intercept** and be careful with your positive and negative numbers. You don't need to use the formula that you learned in the last lesson to calculate the slope. You only need to examine the line and count how many squares it goes up and how many squares it goes over. (Since the slope of a straight line is constant, you can use any section of the line to do this.) Then simplify the resulting fraction as much as possible.

Write equations to go with the graphs for Problems 47 through 53.

47.

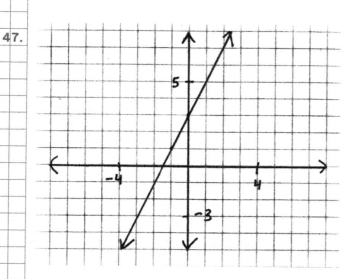

48.

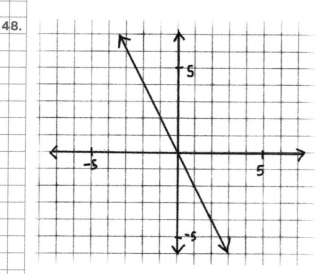

49.

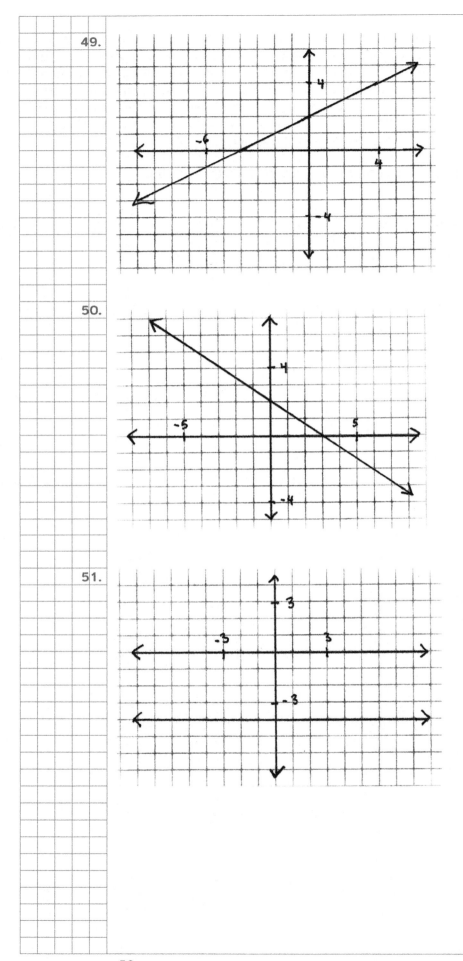

50.

51.

52.

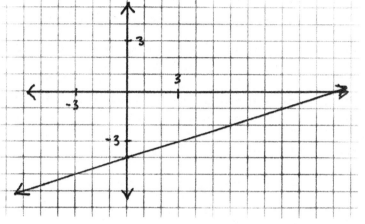

53.

54. It's time for a *Note to Self* about *slope-intercept form*. This Note should explain how slope-intercept form can be used to move easily between the equations and graphs of functions.

REVIEW

1. A group of grandparents and grandchildren are traveling to Vancouver, B.C. using several buses. The ratio of men to women to children is 1 : 2 : 3. If there are 180 people going on the trip, how many men are going? How many women are going? How many children are going?

2. In the following multiplication problem, B represents what mystery digit?

$$\begin{array}{r} 23B \\ \times\ 16 \\ \hline 3,7BB \end{array}$$

3. Emily is able to type 45 words per minute. If she increased her speed by 20%, what would be her new typing speed?

4. In juggling class, a quarter of the students failed the final exam. Of the remaining students, one-third earned an A. What fraction of the class passed the test but scored below an A?

5. In the 3-meter springboard diving competition, Josie earned the following scores on her five dives: 7.4, 6.8, 8.0, 6.4, and 7.6. What was her average score per dive?

6. Within the city of Beaverton there are 129,310 television sets. If there are 1.5 televisions per person, how many people live in Beaverton?

Prime factorize the following numbers:

7. 220
8. 175
9. 363

Rewrite the following expressions using exponents:

10. $7 \cdot 7 \cdot 7 \cdot 5 \cdot 5 \cdot 5$
11. 64
12. $\underbrace{a \cdot a \cdot a \ldots \cdot a}_{b \text{ of them}}$

13. $(x - 1)(x - 1)(x - 1)$

3 TABLES, EQUATIONS, & GRAPHS

For a particular function, moving from the formula to the table is fairly straightforward. All you have to do is plug a value for **x** into the formula to get the corresponding value of **y** (or plug in a value for **y** and get the corresponding value for **x**). For example, if the formula is **y = 3x - 5**, you could plug in the value **1** for **x** and get the value **-2** for **y**.

Complete the following function tables. (You'll notice that a few of these equations have things like x^2 in them. Right now you're perfectly capable of making tables for functions like that — and they *are* functions. You'll learn how to graph this sort of equation later in this book.)

1. y = x + 2

x	0	1	2	3	4
y					

2. y = 2x

x	0	1	2	3	4
y					

3. y = -3x

x	0	1	2	3	4
y					

4. y = -5x + 4

x	0	1	2	3	4
y					

5. y = x - 9

x	0	1	2	3	4
y					

6. $y = \frac{1}{2}x + 3$

x	0	2	4	6	8
y					

7. $y = -x + 8$

x	0	1	2	3	4
y					

8. $y = \dfrac{12}{x}$

x	1	2	3	4
y				

Why didn't I ask you to find the value of y when x was zero?

9. $y = x^2$

x	0	1	2	3	4
y					

10. $y = x^2 - 1$

x	0	1	2	3	4
y					

11. $y = -\dfrac{1}{3}x + 3$

x	-3	0	3	6	9
y					

12. $y = \dfrac{15}{x}$

x	-1	1	3	5
y				

13. $y = -4x$

x	0	1	2	3	4
y					

14. $y = x^2 + x$

x	0	1	2	3	4
y					

In the previous problems, I've been giving you values for **x**. For the following problems, I'd like you to choose your own values for **x**. I'd suggest that you choose those values strategically: you can make things easy for yourself if the **x-values** result in whole-number values for **y**. In Problems 6, 11, and 12, I'm sure you noticed I changed the values I "plugged in" for **x**. Instead of using consecutive numbers, I chose numbers that would yield integer values for **y**.

Set up function tables for your x- and y-values just as you've been doing. Each table should have at least four pairs of values.

15. $y = -x$

16. $y = 3x - 2$

17. $y = -7x$

18. $y = \frac{1}{4}x$

19. $y = \frac{24}{x}$

20. $y = x^2 - 1$

21. $y = -x + 10$

22. $y = x^3$

23. $y = x + 8$

24. $y = -3$

What about moving from table to formula? This is more challenging and (I think) more interesting. First, I'm just going to ask you to throw yourself into trying it for a few problems. Then I'll ask you study to what you've been doing in order to find patterns.

Find the equations that go with the following tables. (It may help to look back at the previous problems...)

25.

x	0	1	2	3	4
y	3	4	5	6	7

26.

x	0	1	2	3	4
y	-7	-6	-5	-4	-3

27.

x	0	1	2	3	4
y	0	3	6	9	12

28.

x	1	2	3	4	5
y	60	30	20	15	12

(Hint: What do you get when you multiply x by y?)

29.

x	0	1	2	3	4
y	1	2	5	10	17

30.

x	0	1	2	3	4
y	6	7	8	9	10

31.

x	0	1	2	3	4
y	-7	-5	-3	-1	1

32.

x	0	1	2	3	4
y	1	6	11	16	21

33.

x	1	2	3	4	6
y	36	18	12	9	6

34.

x	0	1	2	3	4
y	-6	-1	4	9	14

35.

x	0	1	2	3	4
y	9	9	9	9	9

36.

x	3	3	3	3	3
y	0	1	2	3	4

(This may be a little tricky. In a way, it's a lot like Problem 35. It's also the only one that is not technically a function.)

Before we go any further, here's a reminder about vocabulary. In the equation **y = 3x + 5**, the **3** is called the *coefficient* of **x** — it's the number that **x** is multiplied by. It can be positive, negative, or zero. It can be a whole number or a rational number. In that same equation, the **5** is called the *constant*, because it stays the same regardless of how **x** and **y** change.

37. Look back at the previous sets of problems (1-36). Compare those formulae (yes, that's really the proper plural form of "formula," although you can say "formulas," too) that have a positive coefficient of x to those with a negative coefficient of x. How does the sign of the x-coefficient affect the y-values in your table?

38. Now compare only those with a positive coefficient of x. For example, compare Problem 25 to Problem 34. What happens to the y-values in the table as the coefficient of x increases or decreases? (Hint: What I'm trying to get at here is the relationship between the coefficient and the size of the jump in y-value when the x-value goes up by one.)

39. Below I've graphed the functions from Problems 2, 3, and 13. Look at them carefully. Each has a y-intercept of zero. The y-intercept, as you'll recall, is simply the point where the line meets the y-axis.

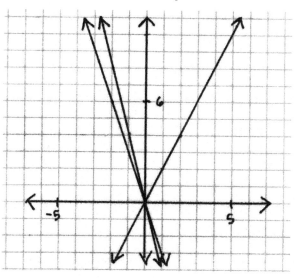

Find two more functions from Problems 1 through 36 that have y-intercepts at zero. How can you recognize from a table that the function has a y-intercept of zero? How do you represent in the equation that the function has a y-intercept of zero?

40. These lines from Problems 1, 5, and 25 have the same slope, but different y-intercepts.

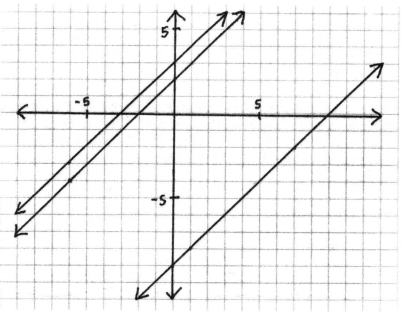

Find two more functions from Problems 1 through 36 that have the same slope and different y-intercepts. How can you tell from the y-values in a table what the slope of the line is? How do you represent that slope when you're writing an equation?

41. Here are the tables and equations for a few of the problems you just did (Problems 3, 4, 7, 8, 12, and 19, if you're curious):

$y = -3x$

x	0	1	2	3	4
y	0	-3	-6	-9	-12

$y = \dfrac{12}{x}$

x	1	2	3	4
y	12	6	4	3

$y = -5x + 4$

x	0	1	2	3	4
y	4	-1	-6	-11	-16

$y = \dfrac{15}{x}$

x	1	3	5	15
y	15	5	3	1

$y = -x + 8$

x	0	1	2	3	4
y	8	7	6	5	4

$y = \dfrac{24}{x}$

x	1	2	3	4
y	24	12	8	6

Notice that for all six functions, as x increases, y decreases. But there are differences in the *way* that y decreases. The first, third, and fifth functions all work one way and the second, fourth, and sixth all work another way. Describe the difference between the two groups in what happens to y as x increases.

Functions such as $y = \dfrac{12}{x}$, $y = \dfrac{15}{x}$, and $y = \dfrac{24}{x}$ are called *inverse functions*.

They have the generic form $y = \dfrac{a}{x} + b$, where **a** and **b** are constants (just as in the form $y = mx + b$).

In the cases of $y = \dfrac{12}{x}$, $y = \dfrac{15}{x}$, and $y = \dfrac{24}{x}$, b is equal to zero, so their generic form could be written more simply as $y = \dfrac{a}{x}$.

42. Solve the equation $y = \frac{a}{x}$ for a.

43. What would you expect the graph of an inverse function to look like? One way to go about answering this question would be to graph the points that you found for the equation $y = \frac{12}{x}$ and then try to figure out what the graph would look like.

Here are some good questions to ask yourself:

- Is the overall shape of the graph going to be a straight line?
- What kinds of values is y going to have when x is very, very big — say 1,000 or 1,000,000?
- What's going to happen to y when x is very, very small — say $\frac{1}{1,000}$ or $\frac{1}{1,000,000}$?
- What is y equal to when x is zero? What does that suggest about the graph?

However you go about it, your job is to sketch what you think the overall shape of the graph of an equation like $y = \frac{12}{x}$ might be.

44. Functions that have an x^2 (or x to any power) are interesting, too. Their y-values increase quite rapidly. What shape would you expect these graphs to have? Again, consider a specific example — say, the function from Problem 9: $y = x^2$. Look at the y-values in your chart for Problem 9.

Do they increase in steady steps? Will the graph of this function be a straight line or a curve? What happens to y when x is zero?

What about when the values for x are negative? Try drawing a sketch of what you think the graph of that function would look like.

(Challenge yourself to avoid the temptation to look at the following page, where the graphs will be revealed.)

On the left are examples of what the graphs of functions with exponents look like; on the right are some graphs of inverse functions.

Parabolic Functions
(not their official name, but it's
 what I like to call them)

Inverse Functions

$y = x^2$

$y = \dfrac{8}{x}$

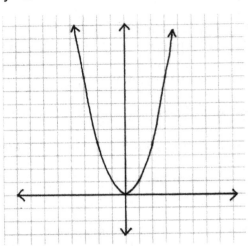 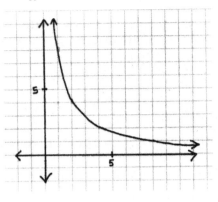

$y = x^2 + 2$

$y = \dfrac{6}{x}$

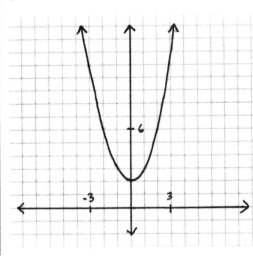 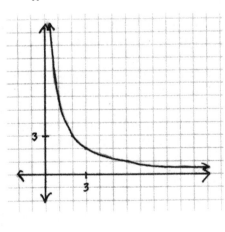

You haven't learned how to actually *make* parabolic or inverse graphs — that will happen in Chapter 4. All I'd like you to be able to do at this point is *recognize* that a function uses an exponent or is inverse when you see its graph, equation, or table.

45. Take a minute to study the graphs that go with parabolic functions. A curve of this shape is called a *parabola*. (I think that this is a rather lovely word for a rather lovely thing.) Explain how the shape of the graph fits with the equation. (Think about the same questions as before: what happens to the y-values as the x-values increase or become negative?)

46. Now look at the graphs of inverse functions. These curves are called *hyperbolas*. One thing to realize about hyperbolas is that (in this form) they never, ever touch the x- or y-axes. How do hyperbolas seem to fit with inverse functions? (What happens to y when x gets very big or very small? I won't ask you to think about what happens when x is negative yet.)

The third type of function that you have been dealing with is called a *linear function*. These are probably the easiest of the three types, and these you do know how to graph.

Linear Functions

y = x   y = -2x + 3

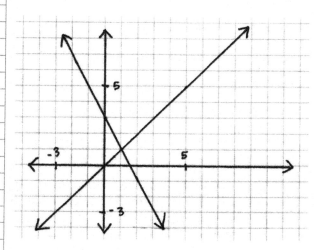

47. It's time for three separate *Notes to Self*: one on **linear equations**, one on *parabolic equations*, and one on **inverse equations**. Each Note should explain how to recognize that a function belongs to that type based on the table, equation, and graph. In the case of linear equations, your Note should also explain how the y-intercept and slope of the graph relate to the equation.

48. Here's a challenge for you. This one is tough, so don't worry if you don't get it. Look back at the examples of graphs of parabolic functions. Notice where the bottom of the parabola touches the graph (this point is called the *vertex* of the parabola) and see if you can figure out how that relates to the equation. The equation for the graph below involves an x^2. See if you can figure out what the equation is. Creating a function table might help you find the relationship between x and y.

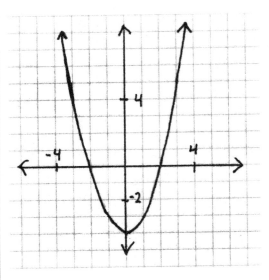

Now it's time to mix it up. I'm going to give you one of the following: a graph, a formula, or a function table. I want you to supply the other two representations of that function.

For example, if I gave you the equation **y = x + 2**, you would supply its table and graph, like this:

x	0	1	2	3	4
y	2	3	4	5	6

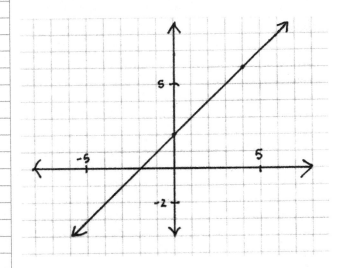

Chapter 2, Lesson 3

Supply the two missing representations:

49. $y = 5x$

50.

x	0	1	2	3	4
y	2	3	4	5	6

51.

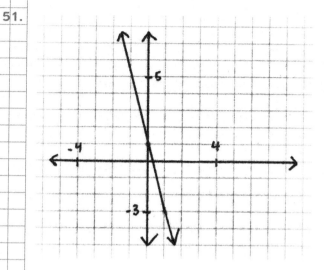

52. $y = -4$

53. Give me the function table, graph, and equation of the function based on the following information: Line AB has a slope of 1/2 and intersects the x-axis at -4.

54. You'll need a small group of partners for this problem, and it will have to be done in school. Your job, as a group, is to visit your teacher and ask him or her to play the Function Game.

REVIEW

Show all work.

1. The Digital Super Store guarantees to refund what you paid for an item plus an additional 10% if you find the item for less at another store. Linus paid $352.90 for a Blu-ray player at the Digital Super Store, then found an ad for the same machine at the Electronics Company for less. How much money should Linus get back?

2. Last year the ratio of boys to girls in Fred's class was 3 : 5. Since that time two additional boys joined the class, no boys left, and the number of girls stayed constant. The ratio of boys to girls is now 4 : 5. How many boys are now in the class?

3. A state-of-the-art alarm system for repelling trolls normally costs $80, but is marked 25% off. With Illinois sales tax of 6.75%, what will be the final price of the troll alarm?

4. A bag of 25 sea slugs costs $1.75. If the bag weighs 5 pounds, find the following:
   a. Dollars/slug
   b. Cents/slug
   c. Ounces/slug
   d. Slugs/pound

5. What is the average of 1/3 and 1/4?

6. If Greg sees an alligator and runs 100 meters in 12 seconds, how fast is he running in kilometers per hour?

7. Imagine an ant weighing 25 grams that can lift a crumb weighing 30 grams. Suppose this ant accidentally falls into a test tube in the lab of a mad scientist and grows into a monster ant weighing 60 kilograms. If its weight-to-strength ratio remains the same, how much can it lift now? (Hint: Think in terms of ratios instead of worrying about unit conversions.) Do you weigh enough to be safe from the monster ant?

Simplify the following fractions:

8. $\dfrac{75mn}{90m}$

9. $\dfrac{32r^{12}}{36r^{8}}$

Simplify the following expressions:

10. $7x + \dfrac{2x}{4}$

11. $\dfrac{2m}{3} - \dfrac{m}{5}$

4 TWO-VARIABLE INEQUALITIES

In *Jousting Armadillos* you practiced working with single-variable inequalities as well as with single-variable equations. It will probably not surprise you to learn that it is also perfectly possible to have a two-variable inequality. And, just as working with single-variable inequalities is very similar to working with single-variable equations, working with two-variable inequalities is very similar to working with two-variable equations.

1. Look back at your Note to Self book from *Jousting Armadillos* and find the Note that has to do with working with single-variable inequalities. What is the key difference between working with single-variable inequalities and with single-variable equations? (Hint: It has to do with multiplying and dividing by negative numbers.)

Here is a quick set of single-variable inequalities to jog your memory. Draw your solutions on number lines.

2. If -x < 3, solve for x.

3. If 42 > -6x, solve for x.

4. If 24 > -4x and -2x > -14, solve for x. Represent x in a single inequality.

5. Suppose 6x ≤ 12 or 2x > 18. Solve each inequality. Can you represent x in a single inequality?

When you work with two-variable inequalities, you need to keep the same thing about multiplying and dividing by negative numbers in mind. Otherwise, manipulating them is exactly like manipulating two-variable equations. Try the following problems, paying attention to what has to be done when you multiply or divide by a negative number.

Solve the following inequalities for y.

6. y + 2 > 5 + x

7. -7y ≥ 14x + 42

8. 2y - 3x < 8

9. x > 3(2y + 4)

10. y + 4 - x < 10

11. 6x - 3y < 9

12. 3y ≥ x + 12

13. 2x < -2y + 2

14. 3x + 2y > 8

15. -3(4x +8y) ≥ 72

16. -5x + y + 7 ≤ 7 + 5x

17. x ≤ 5y + 3

There are a couple of differences between two-variable inequalities and equations when it comes to tables and graphing. First of all, you will very rarely see an inequality turned into a table. To understand why this is, consider the inequality **y > x + 2**. Now, making a table for the related equation **y = x + 2** would be simple. You'd pick a value for **x** and find a corresponding value for **y** that makes the equation true. Your table might, for instance, look like this:

x	y
-1	1
0	2
1	3
2	4

18. **Imagine trying to make a similar table for y > x + 2. If you chose the value 4 for x, what might y be? What else might y be? How many possible values for y are there that go with the value 4 for x?**

So the fundamental problem is that the solutions for two-variable inequalities do not come in pairs. Any value for **x** could have an infinite number of values for **y** that would go with it. (This means, by the way, that an inequality cannot be a function, for the technical reasons that I discussed in Lesson 1 of this chapter, but you don't need to worry about that yet.) You might be tempted to think that since you can't make a table that goes with an inequality, you also can't make a graph of an inequality. In fact, you *can* graph a linear inequality and you'll find that it's not even particularly difficult. To see how it works, think about the following way of putting the inequality **y > x + 2** in a table:

x	y
-1	>1
0	>2
1	>3
2	>4
3	>5
4	>6

You would read this by saying:
"When **x** is negative one, **y** is any number greater than one."
"When **x** is zero, **y** is any number greater than two."
And so on...

In order to make a graph of $y > x + 2$, you have to remember what a graph represents: all of the possible solutions to an equation (or inequality). When you're dealing with an equation, what you're doing is coloring in or darkening the points that represent its solutions — and those points form a single line.

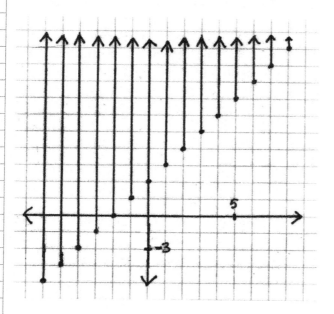

The picture above begins to represent what the table shows us about the graph of possible solutions to the inequality $y > x + 2$. For each **x-value**, the **y-value** can be anything greater than **x + 2**. It's not a complete picture, though, because it doesn't show what happens when **x = 0.5** or **x = 0.25**, and so on. This starts filling in the spaces between points on the graph.

In order to help you figure out how to graph inequalities, I'm going to ask you consider the inequality $y \neq x + 1$. (Remember that this means **y** is *not* equal to **x + 1**.) Now, graphing the equation **y = x + 1** is a piece of cake for you after the last lesson.

19. **Graph the equation y = x + 1.**

20. **Now think about the inequality y ≠ x + 1. The points that make y = x + 1 true are represented by the graph you made for Problem 19. What points make y ≠ x + 1 true? What would a graph of those points look like? (You can describe it or try to draw it if you like, although it's a little difficult to draw.)**

I hope this has given you the idea that you can shade in areas to represent the solutions to inequalities. Graphing something like $y \leq x + 1$ is easier than graphing $y \neq x + 1$. (I also hope you recall that the symbol ≤ means "less than or equal to.")

21. **The first thing to realize is that any point that makes y = x + 1 true will also make y ≤ x + 1 true. Explain why that's so.**

22. This means that in order to graph $y \leq x + 1$, you can start by graphing $y = x + 1$. Go ahead and do that. Now, since any value for y that's less than the values on the line $y = x + 1$ will also make $y \leq x + 1$ true, you're going to shade part of your graph in. I bet you can guess which part. (Hint: The line $y = x + 1$ will form the border between the shaded and the non-shaded parts.) Go ahead and shade it in.

23. In order to test whether you answered Problem 22 correctly, choose three points that lie in the area you shaded and three points that lie in the un-shaded area. If you did Problem 22 correctly, the coordinates of the points in the shaded area should make $y \leq x + 1$ true, and the coordinates of the points in the un-shaded area should make it false. Check to see whether that's the case. If it isn't, go back to Problem 22 and try shading in a different area.

I'd like to point out one thing you've probably already realized: like the lines in your graph, the shaded area is understood to keep going infinitely — *any* **xy** pair on or below that line will make $y \leq x + 1$ true.

Graph the following inequalities. (Solving them for y is probably a good idea, although it's not strictly necessary.)

24. $y \geq x + 3$

25. $y \leq 2x - 4$

26. $y \leq \frac{1x}{3} + 1$

27. $x \geq 2y - 4$

28. $2x + 3y \geq 6$

29. $y \geq \frac{3x}{4} + 3$

Now let's try something like $y > -2x + 2$.

30. **What makes $y > -2x + 2$ different from $y \geq -2x + 2$? How should this make the graphs of the two inequalities different, especially if you think about the line $y = -2x + 2$?**

How to actually show this on a graph is something that you may not be able to figure out for yourself. Mathematicians have chosen to show that the borderline of an area is *not* included by using a dashed line. So the graph of $y > -2x + 2$ would look like this:

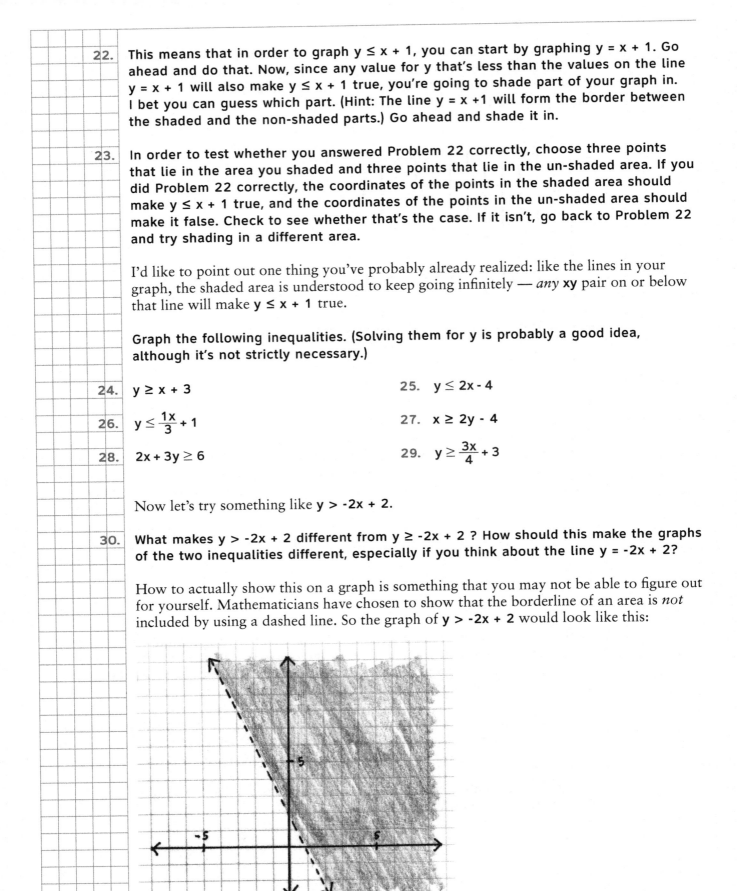

For the following problems, if you're given a graph, give its equation. If you're given an equation, draw the graph. (It will be very helpful to solve the inequalities for y!)

31. $y - 3x > 2x - 4$

32. $2x - 4y \geq 3x + 8$

33. $x - 5y \leq 2x + 10$

34. $x + y < 6$

35.

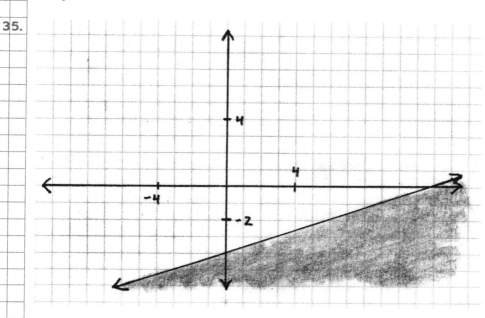

36. $y < 2x + 1$

37. This one might seem tricky, but remember that you've seen this particular curve before. Is this a "greater than" or a "less than" situation? How can you test which one it is?

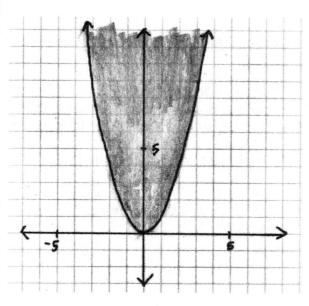

38.

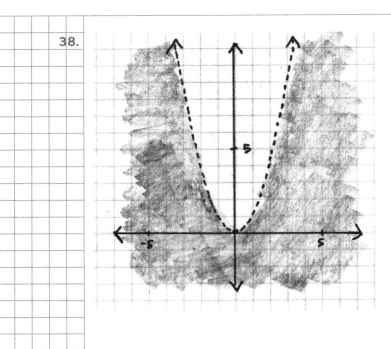

39. Name two points that *are* solutions to y ≥ x² and *aren't* solutions to y > x².

40. Write a **Note to Self** about **two-variable inequalities**. It should include examples of how to manipulate them (including the tricky thing that happens when you multiply or divide by a negative number) and how to graph them.

REVIEW

1. What percent of a foot is an inch? (Round to the nearest whole percent.)

2. Of the 90 members in the scuba diving club, 36 will make dives this weekend. What percent will dive this weekend?

3. In the human body, blood travels at a rate of 4.8 meters in 5 seconds. How far will it travel in two minutes?

4. Matt is driving from Boston to Portland. On Monday he drove 1/7 of the way. On Tuesday he drove 1/2 of the remaining distance. What fraction of the trip does he have left? What percent is that? (Round your answer to the nearest whole percent.)

5. It costs $96 dollars for 5 shirts. How much for 12? Shouldn't you get a discount for buying that many shirts?

6. A room is 1/3 full of people. After 25 people leave, the room is 1/4 full of people. How many people does the room hold?

7. A mother is 28 years older than her daughter. In eight years the mother will be three times as old as her daughter. How old is each now?

8. If the ratio of whooping cranes to sandhill cranes is 1 : 1,050 and the ratio of sandhill cranes to hooded cranes is 50 : 1...

a. What is the ratio of whooping cranes to hooded cranes?
b. If there are 9,975 hooded cranes in the world, how many whooping cranes are there?

9. Simplify: $-3(x-4)+7+2\left(3-\dfrac{1}{4}\right)\div\dfrac{1}{4}-2$

10. Consider the inequality m > x > n. Suppose there is no solution. What does that tell you about m and n?

5 FUNCTIONS AS STORIES

In this lesson, you'll see very few equations (and no inequalities). Instead, you'll look at graphs and, as the title of the lesson suggests, you'll tell stories. Even though you won't be using many equations, this work has everything to do with the sorts of two-variable equations that you've been studying. As you know, every two-variable equation has an infinite number of solutions, those solutions come in pairs, and those solutions can be represented as a graph. So far, you know three basic families of equations: linear, parabolic, and inverse.

I've claimed before that the shift from one-variable to two-variable equations is the most important shift in basic algebra, but so far I don't think that I've done much to convince you or explain why that is. So here's the reason:

The world changes.

All right. You had better be confused by that statement. Let me explain. One-variable equations are very useful if you want to figure out an unknown thing.

For instance, use a single variable to solve this problem:

1. **Giggles and Lucky are two juvenile meerkats. Right now, Giggles is six inches taller than Lucky. Once they've each grown two inches, Giggles will be twice as tall as Lucky. How tall is each meerkat?**

Excellent work. As I say, one-variable equations are quite useful. So what do you now know? You know how tall Giggles and Lucky are at this precise moment. But the thing is that Giggles and Lucky are growing, right? Their lengths are changing, the way that most things in our world change. And two-variable equations, unlike one-variable equations, give us a mathematical way to describe change.

Or, to put it differently, two-variable equations can tell stories.

For the first part of this lesson, the stories I'll ask you to think about and to tell have to do with candles. For centuries, candles were used as clocks, especially at night or on cloudy days. The earliest reference to candle clocks is from Japan around 1,500 years ago, but they were used in many other places and times. The most sophisticated candle clocks were used in the Middle East about 800 years ago: the candles rested in weighted dishes and, as the candles burned away, the changing weights would turn dials.

Here is the story of a candle, told in graphical form:

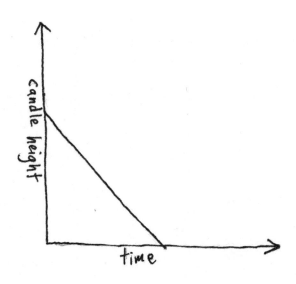

Here's how I might choose to tell the story of that graph in English: "A candle starts off at a certain height. When it is lit, the candle shrinks steadily until it entirely disappears." Notice that, since there are no units on the graph, I can't specify exactly how tall the candle was to start with or how fast it burned, but I know it shrank steadily and disappeared. Also notice that, unlike the graphs you've been working with so far in this chapter, the candle's line does not have arrows at the ends. This means that the candle does not keep shrinking into negative height (whatever that is!).

Here's another version of the candle's story:

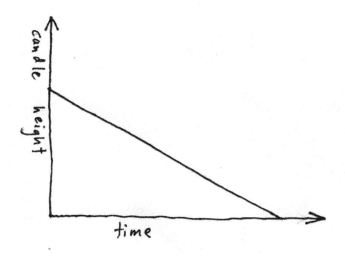

2. **Assume that the mystery units of measurement for the two candle graphs are the same and compare the two stories. What's different about the second story?**

Here's a third version of the story (one that I think is slightly more realistic):

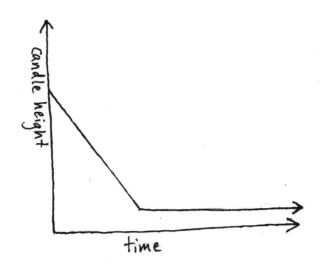

3. **Tell this story. (Remember that the arrow at the end of the candle's graph means something like "continues on like that forever.")**

Tell the version of the candle's story that goes with each of the following graphs:

4.

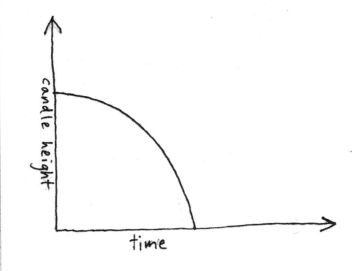

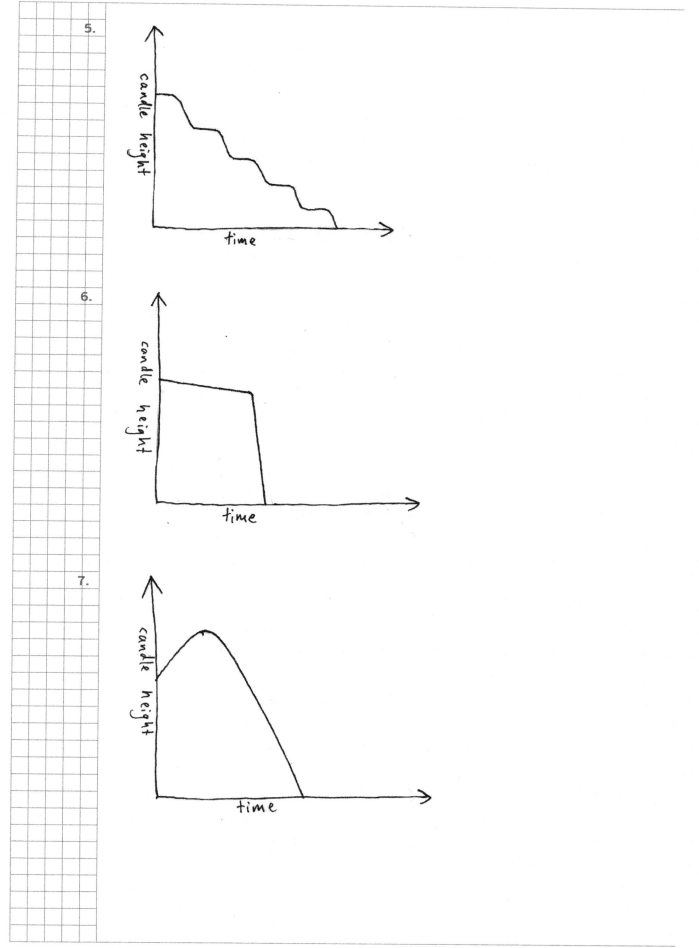

5.

6.

7.

8.

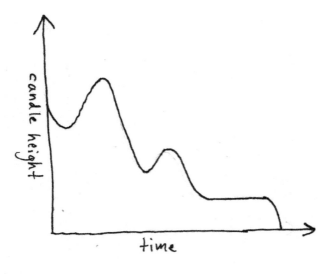

9.

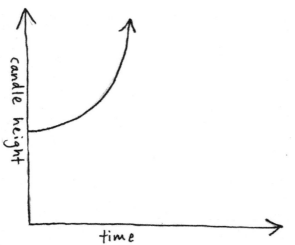

10. **Sketch two candle graphs of your own and tell the stories that go with them.**

Now I'd like to return to the idea of a function, which is what this chapter is all about, after all. In Lesson 1 of this chapter, I told you that a function is "a pairing of two sets of numbers so that to each number in the first set, there corresponds exactly one number in the second set." Then I told you this set of numbers was a function:

x	5	6	52	17	-13
y	7	10	-5	-4	10

... and that this set of numbers wasn't a function:

x	-2	3	3	5	-6
y	12	4	9	16	8

Well, another way to think about this is that, for something to be a function, *each **x-value** can have only one **y-value** that goes with it.*

11. Pause for a moment to answer this question: if a set of numbers had more than one y-value for a single x-value, how could you recognize that fact from the graph of that set? In order to answer this, it might help to graph that second set of numbers above, the set that isn't a function. Pay particular attention to the two points that both have an x-value of 3.

So why does this idea of not having more than one **y-value** per **x-value** matter? Why should that be the defining characteristic of a function? The answer to that question really has to do with stories.

12. In order to try to understand why that definition of a function is important, try to tell the story of this candle graph:

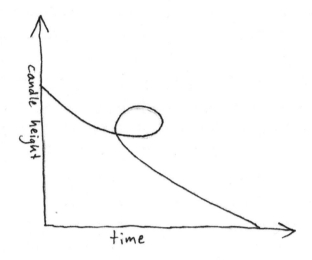

13. I hope you had a hard time with that last one. In fact, you may have found it to be impossible. Copy a sketch of that graph into your notebook and mark the part of the graph that made the story hard to tell.

The graph from Problem 12 does not represent a function because there are certain spots where a single **x-value** (time) corresponds to more than a single **y-value**. Here, for example, I've marked one of those spots with a vertical line:

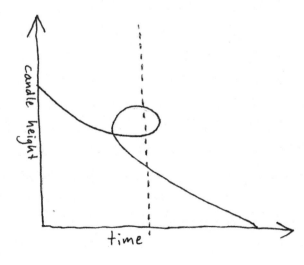

Now, some of the graphs that I asked you to tell stories for earlier were probably impossible. I'm thinking here of the graphs for Problems 7, 8, and 9. Candles pretty rarely grow after you light them. You should be very surprised if you see that happen. (As some of my students have observed, it might be possible to *add* wax to a candle, but Problem 9 would still be pretty difficult.) However, the story that goes with Problem 12 is impossible in a different way, precisely because it's not a function. The line that I put through that graph would have to mean that somehow *the candle was three different heights at the same time.*

That's why mathematicians bothered to come up with the concept of a function. Algebra is useful partly because it can represent aspects of the real world and can be used to make predictions about aspects of the real world. That prediction-making ability is more or less useless if it's possible to have more than one value of, say, candle height for a single value of time. In other words, you can't use an equation to make meaningful predictions unless the equation is a function.

14. **Based on what you now know, explain why an inequality can never be a function.**

Notice that it's absolutely fine for there to be two **x-values** that correspond to a single **y-value**. Here, for example, is a graph of the height of a brick that's been tossed in the air. (This time I have used units on the **y-axis**.)

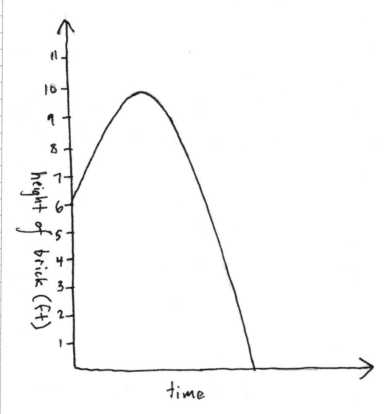

Here's a story that might go with it: "Linus (who's about 6 feet tall), tossed a brick in the air. Then he got the heck out of the way as the brick fell all the way back to the ground." There are a bunch of points on that graph where there are two **x-values** that go with a single **y-value**.

Here's one:

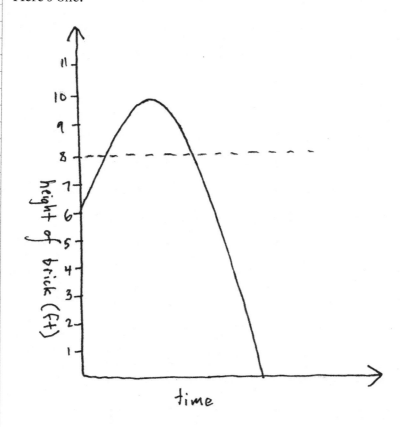

But that's absolutely fine. In fact, it makes perfect sense — there *should* be two different times where the height of the brick is the same: one on the way up and one on the way down. That doesn't ruin the story and it doesn't change the fact that this graph represents a function.

One of the best tests of whether or not a graph represents a function is to mentally run a bunch of *vertical* lines through it — if any one of them would pass through the line of the graph more than once, then the graph isn't a function.

All of the graphs that I've asked you to look at so far have had "time" plotted on the **x-axis**. Although that's true of many functions that you'll encounter, it doesn't have to be the case. For example, you may remember making a graph similar to this one:

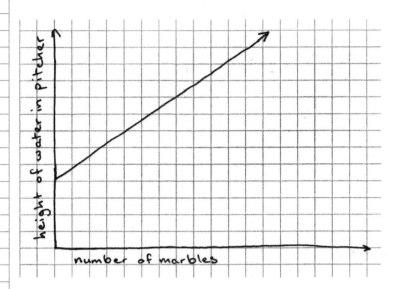

Here's one way of telling the story that goes with that graph: "As you add marbles to the pitcher, the height of the column of water in the pitcher increases at a steady rate. This process can go on forever."

15. **How is the graph below different from the last one and what is the story it tells?**

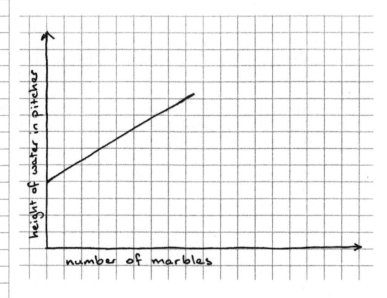

Which version of the story do you think is a more accurate representation of reality and why?

Imagine that the Crow and the Pitcher experiment had turned out differently and make a graph to go with each of the following versions. (Be sure to label your axes each time.)

16. As you add marbles to the pitcher, the height of the water increases. Each marble raises the height of the water by more than the last marble did. This process goes on forever.

17. As you add marbles to the pitcher, the height of the water increases. Each marble raises the height of the water by more than the last marble did. This process goes on until at some point the pitcher breaks and all of the water comes pouring out.

18. As you add marbles to the pitcher, the height of the column of water decreases. Each marble decreases the height by less than the last marble did. This goes on forever and the water never disappears entirely.

For each of the following scenarios, create a graph and write the story that goes with the graph. All of the graphs must be functions, but otherwise the stories can be as creative as you like. No one will be able to read your graphs, though, if you don't label the axes! (For each scenario, I've indicated which part should go on the vertical axis and which on the horizontal.)

19. Linus's height (vertical axis) changes over the course of his lifetime (horizontal axis).

20. The distance a circus performer travels (vertical axis) changes depending on how much gunpowder is in the cannon (horizontal axis).

21. The size of a goldfish (vertical axis) changes depending on how much it eats (horizontal axis).

22. The length of time you spend in jail (vertical axis) changes depending on how much money you give the judge (horizontal axis).

23. How fast you dance (vertical axis) changes depending on how many bees are in your pants (horizontal axis).

REVIEW

Create mathematical expressions to represent the following:

1. Four feet taller than my pet mouse
2. Fifty fewer echidnas than I had yesterday
3. Nine times as many pancakes as Greg can eat

Simplify the following expressions:

4. $|-2 - 6| \cdot |2 - 6|$
5. $3 \cdot |(2)(-5)|$

Find the slopes of the lines that contain these points:

6. (7, 10) and (-1, -6)
7. (-5, 0) and (10, 3)
8. (-6, 10) and (0, -5)

9. Linus cooked a roast beast at 325 °F for 4 hours. The internal temperature rose from 32 °F to 145 °F. What was the average rise in temperature per hour?

10. Here's a laser maze created by my student Izzy Owen.
(Ask your teacher to make a copy of this page that you can work on.)

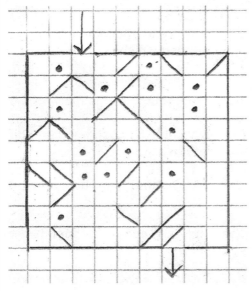

The arrow is a laser beam that enters the maze on the upper left and exits on the lower right. The laser beam travels in a straight line until it hits a mirror (one of the slanted lines), which makes it bounce at 90 degrees, like this:

The beam is required to pass through all of the dots. If the beam runs into a wall, it stops. Fill in the blank squares with mirrors as needed so that the laser beam can escape the maze on the lower right.

6 FORMULAS

I've told you that functions — and hence two-variable equations — can be used to describe change in the real world. Many of the graphs that you created and wrote about in the last lesson are still beyond your ability to translate into equations at this point. Indeed, many of them are beyond my ability (which is definitely *not* the same thing as saying that *no one* can write equations for them). However, there are some graphs that we are both perfectly qualified to translate into equations using the skills we have now. For instance, here's a more precise version of the very first candle story you saw:

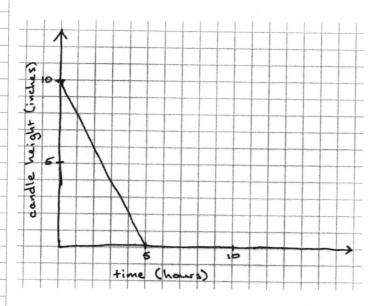

This graph is very, very much like the ones that you were working with throughout the first three lessons of this chapter. So...

1. **Write the equation that goes with that graph. (As you can probably tell, x should stand for time and y for candle height.)**

Here's another version of the candle story:

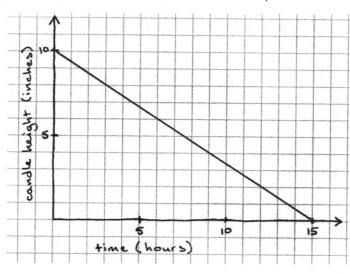

2. Write the equation that goes with that graph.

3. In each of the two graphs that you just worked with and in each of the equations that you wrote for them, what aspect of the candle story does the y-intercept represent? What does the slope represent?

Notice, as you did in Lesson 5, that in neither of those graphs does the line representing the candle's height end in arrows. In other words, the graphs and the equations are only valid from the height at which the candle was lit to the height at which it no longer exists. It's pretty typical for a function to be valid only for a finite set of values. I'll have more to say on this subject in the next lesson.

4. In the equation $y = -6x + 12$, y is a candle's height in inches and x is the time elapsed in hours after the candle is lit. Without making the graph, can you tell how tall the candle is at the moment it's lit? How tall will the candle be after half an hour? How tall will the candle be after 1 hour? At what rate is the candle burning?

5. In the equation $y = 4x + 10$, y is a candle's height in inches and x is the time elapsed in hours after the candle is lit. Just by looking at the equation, is this candle shrinking, or is it a magical growing candle? Again, without making the graph, tell how tall the candle is when it's lit. How tall will it be after it's burned for three hours?

When an equation is used to describe the real world, it's often called a formula. The useful thing about a formula is that you can use it to make predictions. In the case of these formulas, you can figure out the height of the candle at any time: all you need to do is plug the amount of time that has passed into the formula and you get the height of the candle. For instance, you could say that the formula for the candle's height in Problem 1 is $y = -2x + 10$.

Let's look at one more formula from the last chapter. You are not quite ready to write the formula that goes with the story about Linus throwing a brick into the air. You'll study more about this sort of formula in the next chapter. You are qualified to test a claim about this formula that I'm about to make, however.

6. I claim that this formula can be used to describe the situation where I threw the brick in the air: $y = -(x - 2)^2 + 10$. In this case, y is the height of the brick in feet and x is the time that passes in seconds after I throw the brick. In order to see whether this formula corresponds with the graph from the last chapter, make a table for the x-values 0, 1, 2, 3, 4, and 5. Then graph those points (label your axes!) and connect them with a smooth curve. Do you agree with the formula I've proposed?

Scientists and mathematicians use formulas all the time. In the remainder of this lesson, I'm going to ask you to look at three real-world formulas that belong to the three families of functions that you've studied so far. Before you do that, though, you need to learn a couple of pieces of vocabulary. In every formula, there is an *independent variable* and a *dependent variable*.

7. In the case of the candle stories, you had two variables: x was used to represent time and y was used to represent candle height. In these stories, did the candle height depend on the time or did the time depend on the candle height? Based on your answer to this question, in a formula like $y = -2x + 10$, is x the dependent or the independent variable? Is y the dependent or the independent variable?

As I hope you just managed to figure out, when an equation is in **y-equals** form, **x** is the independent variable and **y** is the dependent variable. The height of the candle depends on how much time has passed. The height of the brick depends on how much time has passed. The height of the column of water in the pitcher depends on the number of marbles you've put into it. How fast you can dance depends on the number of bees in your pants. You can also say that **y** *is a function of* **x**. The height of the candle is a function of time. Your dancing speed is a function of the number of bees in your pants.

8. Spend at least five minutes making a brainstorm list. The list should name as many things in the real world as you can think of that depend on each other and that could possibly be described with a number. Here's one to start with: "How fast a soccer ball goes depends on how hard you kick it." (The speed of a soccer ball and the force of a kick can both be described with numbers.) Don't stop until there are at least ten good items on your brainstorm list.

In theory, all of the relationships of dependence that you just brainstormed could be described by formulas, though many of those formulas might be very, very complicated. Formulas are incredibly useful. I think that algebra is beautiful in and of itself, but it is also the foundation of a lot of very practical mathematics. Engineers routinely use algebra for designing and building everything from light bulbs to suspension bridges. Artists and craftspeople use algebra to plan everything from a wool sweater to a wooden cabinet. Algebra is used in every branch of science to describe everything from how the earth orbits the sun to the ways that proteins fold and unfold inside your cells. It is two- (or more) variable equations that make all of these descriptions possible.

As I've said, many of the formulas that are used by engineers, artists, and scientists are very complex, since they describe complex situations. The world is complicated. Even the case of the soccer ball is potentially very complicated, since the speed at which it travels doesn't depend only on how hard it's kicked — it depends on how well inflated the ball is, where exactly your foot makes contact with the ball, how much resistance the ball encounters from the air and from the ground, and so on. In other words, neither of us is ready to start designing suspension bridges just yet. But the skills you've been learning are the foundation for those that you'll need to design suspension bridges and to do many other things. And we are ready at least to begin looking at some real-world situations and the ways that algebra is used to describe them and to make predictions about them.

We'll start with a formula that I bet you've seen before: $E = mc^2$

Albert Einstein, a famous physicist whose photograph I'm certain you've seen, came up with the formula $E = mc^2$ in 1905. The equation describes how much "rest energy" an object contains. According to Einstein (and others), all things contain energy even when they're just sitting still, and this is known as rest energy.

85

Let's look at the formula in detail. The first thing you need to know is that although it looks like it has three variables — E, m, and c — only two of those are actually variables. E stands for rest energy, the dependent variable; m stands for the mass of the object, the independent variable, since the object's rest energy depends on its mass. c, however, represents the speed of light, which is actually a constant.

9. **Can you figure out why a letter is used to represent the speed of light in the formula $E = mc^2$? The speed of light is 299,792,458 meters per second. If you were to multiply 299,792,458 by 299,792,458 (that's c^2, right?), how many rows of partial products would you have to add in order to get the final product? How many digits would the final product have? (Try rounding the speed of light to the nearest hundred million in order to answer this question.) Now, why is it that the speed of light is represented by a letter instead of a number in $E = mc^2$, especially on T-shirts that you sometimes see with that formula?**

As you'll notice, I've used the unit meters per second for c. For m I'll use kilograms, and for E I'll use a unit that you probably haven't encountered before. It's called a *Joule*.

So, if c^2 were translated into a number, the formula would look like this (assuming my calculations are correct — check them if you want to):

$$E = 89,875,517,873,681,764m$$

Notice that in this version I flipped the m and the value for c^2. That's just because in a situation like this one, where two things are being multiplied, I like to put numbers first and variables second.

10. **You know three "families" of functions. The formula $E = 89,875,517,873,681,764m$ belongs to one of those families. Which one does it belong to and why? (One way of thinking about this is to ask the question, "What's happening to the independent variable in this formula?") Sketch the graph of this formula (a sketch only, mind you: an accurate graph would require either an absurd scale or a very large sheet of paper).**

11. **What would the slope of the graph of $E = 89,875,517,873,681,764m$ be? What would its y-intercept (or, in this case, its "E-intercept") be?**

So the formula $E = 89,875,517,873,681,764m$ belongs to the linear family of functions. Not only does the rest energy of an object increase as the mass of the object increases, but it increases in a regular fashion: each additional kilogram of mass adds 89,875,517,873,681,764 Joules of rest energy.

12. **What is Linus's approximate rest energy if his mass is approximately 100 kilograms? (It's actually 91 kilograms.)**

Now, a Joule is not a very large unit of energy; it's about the amount of energy that's released when a heavy book is dropped on the floor. But you can see that if the rest energy contained in Linus's body were somehow released, it would be a very large amount indeed. (Imagine that many heavy books being dropped on the floor at the same time.)

In fact, partly through the work that Einstein did, scientists have figured out how to release some of the energy contained in the tiny nuclei of atoms in the form of nuclear power (and nuclear bombs).

$E = mc^2$ is one example of *linear variation* (or *direct variation*). There are a couple of other ways to say this. You can say that E *varies in a linear fashion with* m, or that E *varies directly with* m.

Let's look at a few other examples of linear variation that you're familiar with, although you probably didn't realize before that they were examples of linear variation.

Here is the formula for finding the perimeter of a square from the length of one of its sides: $P = 4s$

... where P stands for perimeter and s stands for the length of a side. You know perfectly well that the perimeter of a square is four times the length of one of its sides, although you may never have written out the formula before.

13. **Make a table showing the perimeters of four different squares with side lengths of 1, 2, 3, and 4 respectively. ("1, 2, 3, and 4 *what*?" you may ask. Because perimeter and side length are measured in the same units, the unit doesn't matter. A square with a 1-inch side has a 4-inch perimeter. A square with a 1-mile side has a 4-mile perimeter.) Now make a graph of the data in your table. Because a square can have a side length that's between 1 and 2 (or 2 and 3, and so on), you can connect your points with a line and, because the squares can keep getting bigger infinitely (at least in theory), you can extend that line and put an arrow at its end. What is the slope of the line? What's its y-intercept (or "P-intercept")?**

So you can say that the perimeter of a square varies directly with the length of its sides (or that the perimeter varies in a linear fashion with the lengths of its sides, or that this is an example of linear variation).

14. **You can extend the line in Problem 13 toward infinity in the up-and-right direction, but it needs to stop at the origin in the down-and-left direction. Why?**

The thing that you noticed in Problem 14 is true about many formulas, making them unlike the graphs that you studied and created in earlier lessons. It doesn't really make sense to have a negative square. You can't have negative rest energy either (unless there's such a thing as negative mass — but we won't get into the concept of antimatter here).

One more example of linear variation — the formula for finding the circumference of a circle: $C = 2\pi r$

C is the circumference and r is the radius in this case.

15. **What is the circumference of a circle that has a radius of 1, 2, 3, and 4? (Just as in Problem 13, you don't need units here.) If you were to graph the relationship between circumference and radius, what would the slope of the line be? What would its y-intercept (or "C-intercept") be?**

Now let's take a look at another type of variation that belongs to another family of function that you already know. In this case, we'll use another, less famous formula that also has to do with energy. This time the energy involved is called *kinetic energy*, which is the energy of motion. The formula for kinetic energy is $K = \frac{1}{2}mv^2$. **K** stands for kinetic energy, measured in Joules again; **m** stands for mass, just as it did in the last formula; and **v** stands for velocity, which means speed and is measured in meters per second. In this formula, unlike the last one, there are three genuine variables. **K** is the dependent variable, but **m** and **v** are both independent variables; kinetic energy depends on the mass of an object and on its velocity. We're not really ready to deal with a three-variable equation yet, so we're going to imagine how the kinetic energy of a single object changes when the velocity of that object changes. The object that we're going to imagine is Greg, whose mass is about **86** kilograms. (I find it hard to imagine that Greg's mass is really smaller than mine. He's at least two inches taller and a lot more muscular. One of us is either wrong or lying, but we'll just have to accept for now that Greg's mass is **86** kg.) Having one variable stay the same throughout a problem is something mathematicians do all the time; it's called *holding that variable constant*.

16. Make a table of the values of v and K that answers the following questions. What is Greg's kinetic energy when he's standing still? (That's v = 0.) What's his kinetic energy when he's hopping forward at 1 meter per second? He increases his hopping speed to 2 meters per second; what's his kinetic energy now? How about at 3 meters per second? Four meters per second? What if he's sprinting at 10 meters per second? How about riding the Linuscycle at 100 meters per second?

17. Before you make the graph, in the case of $K = \frac{1}{2}mv^2$, is K the dependent or the independent variable? How about v? Therefore, which one should be on which axis of the graph you're going to make?

18. Looking at the values in your table from Problem 13 and at the equation $K = \frac{1}{2}mv^2$, what do you expect the shape of the graph to be? Test your hypothesis by graphing the first five points from the table and connecting them with a smooth line. Even using only those five points, you'll need a highly modified scale. I recommend this: four graph paper squares equal 100 Joules on the vertical axis and four graph paper squares equal 1 meter per second on the horizontal axis. Even at that scale, expect to use basically a whole page of graph paper for this graph. Label your axes!

19. You've seen that shape before, of course, in Lesson 4 of this chapter. What's the name of the shape of that graph? What feature of the equation $K = \frac{1}{2}mv^2$ should lead you to expect a graph of that shape?

20. $E = mc^2$ has a straight line graph and $K = \frac{1}{2}mv^2$ has a parabolic graph (when you hold m constant) even though both formulas contain exponents. Why is that the case?

So the formula $K = \frac{1}{2}mv^2$ belongs to the family of parabolas; the other way to say this is that **K** *varies with the square of* **v**.

We'll look at two (probably more familiar) examples of varying with the square. Here's the formula for finding the area of a square from the length of one of its sides: $A = s^2$

... where **A** is area and **s** is the length of a side.

21. What is the area of a square with a side length of 1 inch? How about 3 inches? How about 8 inches? Draw a sketch of what the graph would look like — just get its overall shape without worrying about the exact points.

22. In Problem 21, I hope you drew only half of a parabola. Why doesn't it make sense to draw the left half?

The formula for the area of a circle is $A = \pi r^2$. You can probably figure out that A is the area and r is the radius.

23. If you have a square with a side length of 4 miles and a circle with radius 4 miles, which one has a greater area? How many times greater? Keeping this fact in mind, on the same graph where you drew a sketch of the relationship between a square's side length and its area, draw a sketch of the relationship between a circle's radius and its area. (Be sure to label which is which.)

Let's take a look at one more formula that belongs to one other family — the last family that you're familiar with. In this case, the formula is:

$$D = \frac{m}{v}$$

This formula, like $K = \frac{1}{2}mv^2$, really does have three variables in it. Once again, m stands for mass; this time we'll measure it in grams. In this case, v stands for volume (that's how much space an object takes up), measured in cubic centimeters. D stands for density (how tightly packed the material in an object is — a brick is quite dense, while a block of packing foam is not very dense at all), and is measured in grams per cubic centimeter.

Just as we did with $K = \frac{1}{2}mv^2$, we'll deal with the fact that $D = \frac{m}{v}$ has three variables by holding one of the variables constant.

Imagine that you have a tied-off balloon attached to a special machine that can stretch the balloon out or squeeze the balloon without putting more air into it or taking any air out.

24. Carefully consider the scenario I've just described. Which variable is being held constant (it doesn't change when the machine is operated): density, mass, or volume?

25. In this scenario, of the two variables remaining, which one is the independent variable and which is the dependent? What does that imply about the axes of the graph you're going to make?

26. The mass of the balloon is 240 grams. Make a table to answer the following questions: When the volume of the balloon is 1 cubic centimeter, what's its density? What's its density when it's stretched to 2 cubic centimeters? How about 3? 4? 5? 10? 12? 20? What if the machine compresses it from its original 1 cubic centimeter to half a cubic centimeter? What about a third of a cubic centimeter?

27. As the volume of the balloon increases, does the density increase or decrease? Does the increase or decrease happen in regular intervals? (In other words, is it linear?) What shape do you expect the graph of this function to have?

28. Test your hypothesis from Problem 27 by graphing some points from the table in Problem 26. Graph the values for v = ½, 1, 2, 3, 4, and 5. Once again, you'll need to play with the scales of this graph; I recommend a scale of 4 graph paper squares to 100 grams per cubic centimeter on the vertical axis and 4 graph paper squares to 1 cubic centimeter on the horizontal axis. Once again, you'll need most of a graph paper page for this, and, once again, label your axes.

29. In Problem 26, I didn't ask you to calculate the density when the volume was zero. Explain why it would have been mathematically impossible to do that. Thinking about the scenario of the balloon, explain why it would be physically impossible for the volume to be equal to zero. Explain how the graph you made reflects the fact that it would be impossible for the volume to be equal to zero.

30. What's the name for the shape of the graph that you made in Problem 28?

So, the relationship between the balloon's density and its volume clearly falls into the inverse category. Another way of saying this is that **D** *varies inversely with* **v**. I won't ask you to look at any other examples of inverse variation for right now, partly because there aren't any clear examples having to do with squares and circles.

31. As I pointed out earlier, K = ½mv² and D = $\frac{m}{v}$ actually have three variables apiece.

It's quite correct to say that K varies with the square of v and D varies inversely with v. However, K does not vary with the square of m and D does not vary inversely with m. Copy the following sentences and fill in the blanks:

In the formula K = ½mv², K varies _____ with m.

In the formula D = $\frac{m}{v}$, D varies _____ with m.

(If you have trouble answering this question, try holding v constant at 1 in each equation and test what happens to the value of K and D as m increases.)

32. Write a *Note to Self* about *formulas and types of variation.* (You now know three: direct variation, variation with the square, and inverse variation.) Explain what a formula is and what it's for and define the phrases that go with each of the three types of variation that you now know about.

REVIEW

1. The ratio of Greg's woolly hats to Fred's is 7 : 3. Greg has 16 more hats than Fred does. If he gives one-fourth of his hats to Fred, what will be the new ratio of Greg's hats to Fred's?

2. A column 1,976 feet tall is painted red, white, and blue in the ratio 7 : 5 : 1. What length of the pole is painted white?

3. Alison, Brian, and Christina have 285 marbles altogether. The ratio of Alison's marbles to Brian's marbles is 4 : 3. Brian has 15 more marbles than Christina does. How many marbles does Alison have?

4. Often it is difficult to figure out how to approach a ratio problem. One tactic that works well for getting started is to list all your known information and then construct an illustration. For one of the last three problems, carefully lay out the information and pair it with illustrations that could assist you in seeing the problem in a more concrete way. Think to yourself, "How would I illustrate this problem if I had to explain the procedures to a younger student?"

5. Find the Greatest Common Factor (GCF) of the following numbers:

   8 and 15   24 and 56

6. Divide. Check your answers using multiplication.

   5.6 ÷ 1.4   7.2 ÷ 18

7. A straight line passes through the three points (3, -4), (5, 1), and (7, y). What is the value of y?

8. The HMS Armadillo and the Bonnie Woodchuck are similar enough in design that if they race under the same conditions with crews of equal skill, the boat with the larger sail area will have the advantage. Which boat do you predict will win? How much greater is its sail area?

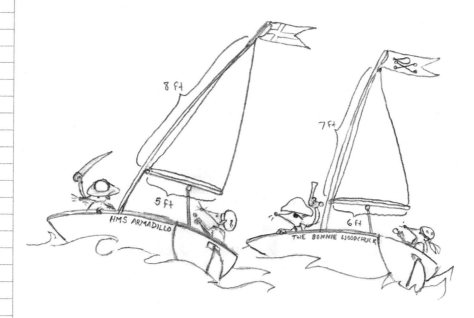

9. This is a tougher version of the Giggles and Lucky kind of problem from the last lesson. As a reminder, I've started a grid to show you one approach to the problem.

The sum of Hans's and Joachim's ages is 45. In five years, Joachim will be sixth-fifths of Hans's age. How old are they?

	Now	Five years from now
Hans	x	
Joachim		

10. This puzzle is adapted from one at www.brainden.com.

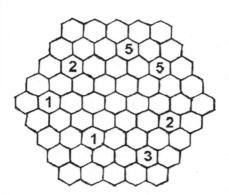

A worker bee makes a looped path (a path that connects back to itself) through the nursery to feed the baby bee in each cell of the comb. His path proceeds from one cell to an adjacent cell through the center of each cell, does not pass through any cell more than once, and never makes an acute-angle turn (i.e., a turn at a 60-degree angle). There are no baby bees in the numbered cells, so he doesn't visit those, but each number indicates how many of the adjacent cells are part of the path. See if you can trace the worker bee's route.

7 INTERPOLATING & EXTRAPOLATING

I hope I managed to convince you in the last lesson that formulas can be used to describe the world and to make predictions about it. If you know the formula for something, you can plug in a value for the independent variable and get the corresponding value for the dependent variable. So how do people come up with these extremely useful formulas in the first place?

Most formulas, at least when they are first being investigated, are based on experimental evidence. In other words, scientists do exactly what you did in the Crow and the Pitcher: they set up an experiment and they try actual values of the independent variable and record what values they get for the dependent variable. (In the case of the Crow and the Pitcher, you added marbles and measured the resulting height of the column of water.)

So oftentimes what scientists have to work with is a set of data rather than a formula.

Here's the set of data from an imaginary (but realistic) experiment. In this imaginary experiment, I've put an imaginary mystery liquid in an imaginary container over an imaginary flame. Then I've put an imaginary thermometer in that liquid to measure its temperature at regular time intervals. Here's what I've found:

Time (minutes)	Temperature (degrees Celsius)
1	31
2	40
3	52
4	61
5	71

1. Plot my data points on a graph. Time is the independent variable and temperature is the dependent variable. Make sure the graph is big enough to work with — I used a scale of 2 graph paper squares to 10 degrees Celsius on the vertical axis and 2 graph paper squares to 1 minute on the horizontal axis. Make sure both of your axes start at zero. And make sure your vertical axis goes up to 110 degrees.

2. You have only five points here, but they're pretty suggestive of a line, right? Of the three kinds of variation that you're familiar with, which one do these data points appear to represent?

3. Go ahead and connect those points in the way that seems most reasonable, extending the line outward in both directions.

What you've just done is called *constructing a line of best fit*. The points don't actually make a straight line, but they're very close. Scientists and statisticians (people who work with statistics) make lines of best fit all the time. There are some very precise mathematical methods that they use to do it. When you take a statistics class, you'll learn about those methods. For now, just drawing the line that you think fits is absolutely fine.

4. **Write an equation to represent the line that you've just drawn. You can go ahead and use x for time and y for temperature.**

5. **Based on the line of best fit that you drew, or on your equation, or on the original data points, or on all three, predict what the temperature of the mystery liquid would have been if I'd measured it at 2 ½ minutes, 3 ½ minutes, and 4 ½ minutes.**

What you just did when you answered Problem 5 is called *interpolating*. To *interpolate* is to make an educated guess about the value of a variable that lies *between* two known values of the variable. You know the temperature at **2** minutes, you know the temperature at **3** minutes, you make an educated guess about the temperature at **2 ½** minutes. In your daily life, you do this kind of educated guessing all the time, although you may not have known before that you were interpolating.

6. **Based on the line of best fit that you drew, or on your equation, or on the original data points, or on all three, predict what the temperature of the mystery liquid would have been if I'd measured it at 0 minutes. Do the same thing for 6 minutes, 8 minutes, 10 minutes, 20 minutes, and 100 minutes.**

What you just did when you answered Problem 6 was called *extrapolating*. To *extrapolate* is to make an educated guess about the value of a variable that lies *beyond* the known values of the variable. You knew the temperature at **1** minute, **2** minutes, **3** minutes, **4** minutes, and **5** minutes. When you guessed the temperature at **0** minutes, you were extrapolating into the past. When you predicted the temperature at **6, 8, 10,** and **100** minutes, you were extrapolating into the future. Just so you know, you're much likelier to hear the word *extrapolate* than the word *interpolate* in ordinary conversation.

Interpolation and extrapolation are both incredibly useful intellectual tools; in essence, they are both forms of inductive reasoning. As you might expect, they do have limitations.

7. **Look back at the values that you predicted in Problems 5 and 6. Which ones do you feel most confident about? Which ones do you feel least confident about? Why?**

Now look at the expanded data table for my imaginary experiment. I've added the value for **0** minutes along with several other values:

Time (minutes)	Temperature (degrees Celsius)
0	20
1	31
2	40
3	52
4	61
5	71
6	82
7	92
8	100
9	100
10	100

8. **Add the new data points to your original graph. Do they conform to your predictions from Problem 6? If not, was it your interpolating or your extrapolating that went awry?**

(My editor, Sarah, would like to take this opportunity to tell you that *awry* rhymes with *deny*. When she was your age she'd only seen it in print and assumed it was pronounced AW-ree. And if you've never seen that word at all, if something is *awry* it's off course or out of the normal position.)

As I've said, although this experiment is imaginary, the data — including the new points I added — are perfectly realistic if the mystery liquid is ordinary water. (In fact, you'd get similar results for any liquid, although the actual temperatures would be different.) If water starts off at room temperature (around 20 degrees Celsius) and is put over a flame, its temperature will steadily increase (surprise, surprise), but that steady increase will happen only up to about 100 degrees Celsius. (The actual temperature at which it levels off varies with, among other things, height above sea level.) At that point, the energy that the flame is adding to the water no longer raises the temperature of the water; instead that energy begins to break the bonds that hold the individual water molecules together. In other words, the water boils. While it's boiling, its temperature holds steady. At some point the temperature does begin to rise again, but it's difficult to measure with an ordinary thermometer because the water has turned to steam.

I'm telling you this stuff about boiling water partly because I think it's interesting in and of itself, but also to make a point about the limitations of interpolation and extrapolation. Interpolation and extrapolation can lead to very good models of the world. After all, the equation you came up with in Problem 4 (I imagine it was something like **y = 10x + 20**) did a pretty good job of representing the experiment up until **x** was equal to **8**. But interpolation and extrapolation can lead to false conclusions. When **x** was greater than or equal to **8**, your equation didn't represent things so well anymore.

This sort of situation arises all the time for scientists; it's often found that a mathematical description is not perfect and needs to be modified (and then modified again, and again, and so on). Indeed, part of the reason that Einstein is so famous is that he figured out that some of the math that had been used to describe the world for literally hundreds of years didn't actually work, especially in extreme situations involving very large or very small objects.

9. Most people would probably say that extrapolating is a little riskier than interpolating, but it's also possible for interpolating to lead to a false conclusion. Here, for example, is a version of the graph that you made in Problem 1 with three new data points added. In this case, the line of best fit that you made at first would have led to incorrect interpolation. (These are definitely not realistic results, by the way!) Tell the story of this graph as it now appears, just as you did for the graphs in Lesson 5.

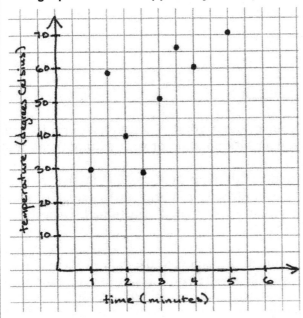

10. Here is a graph of data points from a different experiment. Copy it into your notebook and make a line of best fit for the data points. How confident do you feel that this line would lead to accurate interpolation and extrapolation? Why?

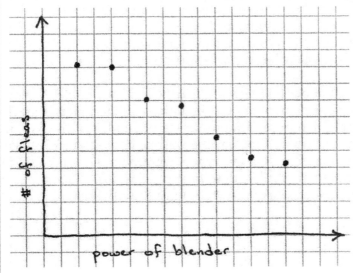

11. Here's another set of data points. Copy the graph into your notebook. To what family of variation does this set of data appear to belong? Draw your line of best fit. How confident are you that this line will lead to correct interpolation and extrapolation? Why?

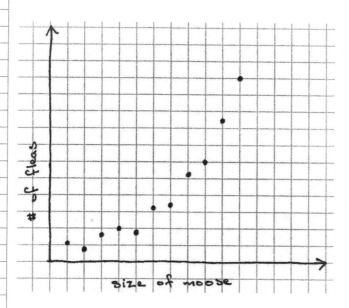

12. Here's another set of data points. To what family of variation does it appear to belong? Copy it and make the line of best fit. How confident are you that this line will lead to correct interpolation and extrapolation? Why?

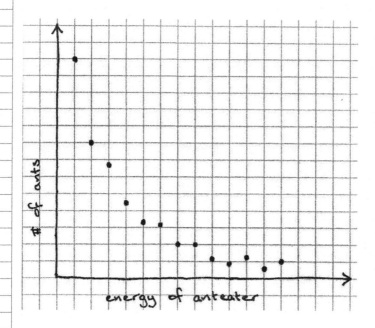

13. Here's another set. Copy it and make the line of best fit. How confident are you this time that it will lead to correct interpolation and extrapolation? Why?

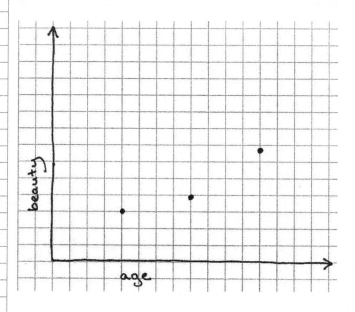

14. Here's another. Copy it, make a line of best fit, and answer the same questions you've been answering.

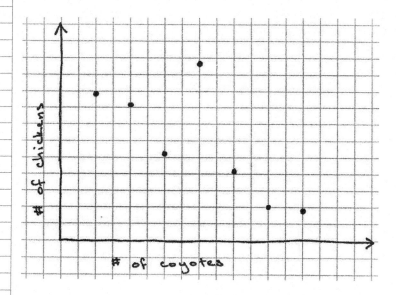

15. One more. Same instructions.

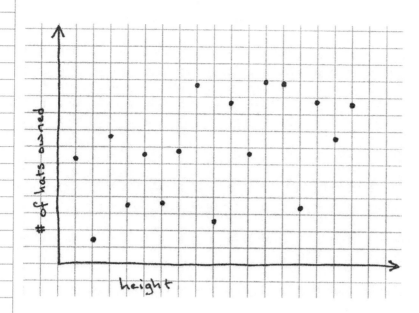

16. According to the chart below, the average American home size has been steadily increasing since the 1940s. Whatever the reasons, this is true. Graph the data points — you'll need a fairly big graph. Your x-axis should be time, and I'd suggest that you call the year 1900 Year 0 and count up from there. (I think calendars are a fascinating subject, by the way. Even though 1900 isn't 0, the fact is that you can start your graph wherever you choose. For instance, according to the Gregorian calendar — the one you're probably used to using — I'm writing this in the year 2010. But according to the Hebrew calendar I'm writing it in the year 5771, and in the Islamic calendar it's the year 1431. Obviously these three calendars use different points as Year 0. The reason for making 1900 be Year 0 for your graph isn't religious, it's practical: it will make it much easier to write an equation for your line.) One graph paper square to 10 years ought to work pretty well for the horizontal scale. For the vertical axis, I'd suggest a scale of 1 graph paper square to 100 square feet, and make sure that the graph goes up to about 3,000 square feet. This graph will take up a whole page in your notebook.

Year	1940	1950	1960	1970	1980	1990	2000
Home Size (ft²)	800	1,000	1,200	1,400	1,600	1,800	2,000

17. Make a line of best fit for the points on your graph. (It will be very easy and accurate in this case.) Then write an equation for the line.

18. According to your graph/equation, how big was the average American home in 1965? 1982? 1997? Did you extrapolate or interpolate to find those values?

Chapter 2, Lesson 7

19. According to your graph/equation, how big was the average American home in 1930? How big will it be in 2030? In 2056? Did you extrapolate or interpolate to find those values?

20. If you decide that you have 100% faith in the accuracy of your line of best fit, it will lead you to some curious conclusions about the past and the future. What does your line of best fit suggest about the living situation of an average American in 1900? What about in 1890? What does it suggest about the average American house in the year 2900? Taken to its logical extreme, what does it suggest about the ultimate fate of the average American house?

21. Obviously, you need to be sensible about how you interpret data and how you apply equations to that data. Up to what points in the past and future do you think your line probably reflects average American home size with reasonable accuracy? Why?

22. Draw a sketch of the graph that you think might actually reflect average American home size from 1800 to 2100.

23. Assuming you didn't think that the average American home was nonexistent in 1900, what do you think might explain the steady increase that began to happen sometime between 1900 and 1940?

24. Assuming you don't think that the average American home will be infinitely big at some time in the future, what do you think might explain the leveling off of change in home size that will have to happen?

25. Write a *Note to Self* about *interpolation and extrapolation*. Define each one and give examples of each. You might also want to point out the limits of interpolation and extrapolation to your future self.

REVIEW

Solve the following equations:

1. $\frac{2n}{3} - 8 = -4$

2. $\frac{m}{2} = \frac{-m}{4} - 3$

Graph the following equations:

3. $y = -x + 5$

4. $2y + 8x = 6$

Graph the following inequalities:

5. $y < 3x - 2$

6. $2x - y \le -5$

7. Write an equation to go with this graph:

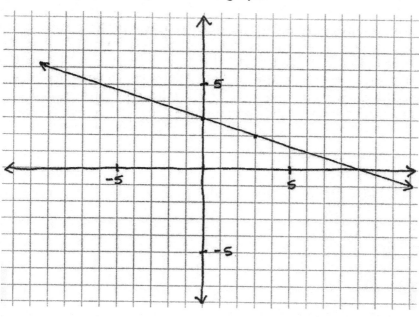

Write equations to go with the following tables:

8.

x	1	2	3	4
y	7	9	11	13

9.

x	0	1	2	3
y	4	5	8	13

10. David and Mary are collecting pumpkins to lob across a field with the giant trebuchet they have built. Last month David collected 3 pumpkins for every 5 that Mary collected. This month they gathered an additional 121 together (not necessarily in the same ratio). David now has three times as many pumpkins as he had last month while Mary has two times as many as she had last month. How many pumpkins did they have last month?

11. Another puzzle from brainden.com:

The Loch Ness monster is lurking beneath the water in the picture below. Nessie is 45 meters long; only three meters of him (or her?) are visible above the surface. His head is at the square labeled "1," the tip of his tail is at "45," and a bit of his middle (the 23rd meter, in fact) is at "23." The black squares are rocks he is curled around and the numbers on the margins show how many squares are occupied by Nessie's submerged body in the corresponding row or column. Where is Nessie?

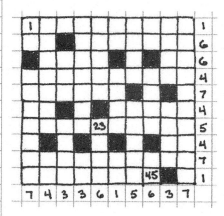

3

SIMULTANEOUS EQUATIONS

1 SOLVING BY GRAPHING

1. There was once a pair of lizards who went everywhere together. Their names, as it happens, were Florence and Edna. In some ways they were quite different: the sum of their lengths was 56 centimeters and the difference between their lengths was 22 centimeters. How long was each lizard?

Now, you could definitely solve this problem by guessing and checking. But this chapter is all about using algebra to solve problems like this, so I am going to ask you to tackle this problem with whatever strategy makes sense to you, but NOT to guess and check. Spend at least five minutes working on it (unless you find a solution faster than that). Here are a couple of suggestions: Try drawing pictures and/or writing equations that describe the lengths of these two lizards.

While working on that problem you may very well have discovered some of the techniques that you'll learn in the next three lessons on your own. Nevertheless, I'm going to explain them to you in detail and give you plenty of chances to practice.

The first step to dealing with a problem like Problem 1 algebraically is to realize that it implies two separate equations. Those equations could be **F + E = 56** and **F - E = 22**. (I'm assuming here that Florence is the longer lizard. You could use **E + F = 56** and **E - F = 22** if you wanted Edna to be the longer lizard. They don't really answer to their names anyway unless you offer them tasty treats.) Usually when you've dealt with different equations, you've assumed that they have different solutions, even when they used the same variables. You must have solved hundreds of single-variable equations that all used the variable **x**, and in each case **x** stood for something new. The key to solving problems like this one is to realize that in these equations, **F + E = 56** and **F - E = 22**, both **F**'s stand for the same number and both **E**'s stand for the same number. The two equations are true *at the same time* and are therefore called *simultaneous equations*.

In this lesson and the next two, you're going to learn three methods for solving simultaneous equations. First we'll look at solving them by graphing. In order to understand why this method works, you have to keep in mind what a graph is: a graph of an equation is a way of representing the infinite solutions that the equation has.

2. The best ways to make a graph from an equation are to find two pairs of solutions that make the equation true *or* to convert the equation into slope-intercept form and use the slope and the y-intercept to graph it. Let's change the lizard problem variables to the more generic y + x = 56 and y - x = 22 so they won't confuse you on the graph. Use whichever method you prefer to graph the equation y + x = 56. (The trick here is that this particular graph will have to be very big. All of the action will be happening in the first quadrant, so that's all you need to draw in this case, but the y-axis needs to go up to 60 and the x-axis needs to go out to 30. I promise that you won't have to make this big a graph for the rest of the problems I'll give you.) For this graph, I recommend using a straightedge to make your lines.

3. Use whichever method you like to graph the equation y - x = 22 *on the same graph*. (The two equations are simultaneous, after all.)

4. The first line you drew represents the infinite pairs of numbers that make the equation $y + x = 56$ true. (If you like to think of it this way, it represents all of the possible pairs of lizards the sum of whose lengths is 56 cm. One lizard could be 55 cm long and the other 1 cm long, one lizard could be 54 cm long and the other 2 cm long, and so on.) The second line you drew represents the infinite pairs of numbers that make the equation $y - x = 22$ true. Where on your graph is the point representing the pair of numbers that makes both equations true?

5. What are the coordinates of that point, or, in other words, what are the two numbers? (Depending on how accurately you drew your graphs, you may have to estimate slightly. Try rounding to the nearest whole numbers.)

6. Check to see that those two numbers really are the lengths of Florence and Edna by finding out whether they make both equations true. (It should work, so if it's off by just a little bit, that probably means that your lines aren't perfectly straight and you can adjust your estimate of their crossing point. If the pair of numbers is off by a lot, it probably means that you actually graphed one or more of the lines *incorrectly* rather than just imperfectly, and you should go back and check your graphing.)

It's actually less important that your graphs in those last problems were perfect than that you understand the principle that the point where the two lines cross each other represents the shared solution to the equations. In the following exercises, you won't have to make graphs that big. However, the shared solution to the equations may lie in quadrants other than the first one. My suggestion is that you leave yourself some room on all sides of a graph the first time you make it so that you can extend it a little bit in any direction if you need to.

Find the point that the two lines share, making your own graphs when necessary. If you're given graphs or tables, write the equations that go with them.

7.

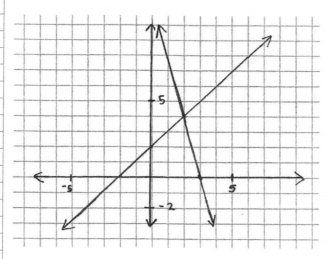

8.

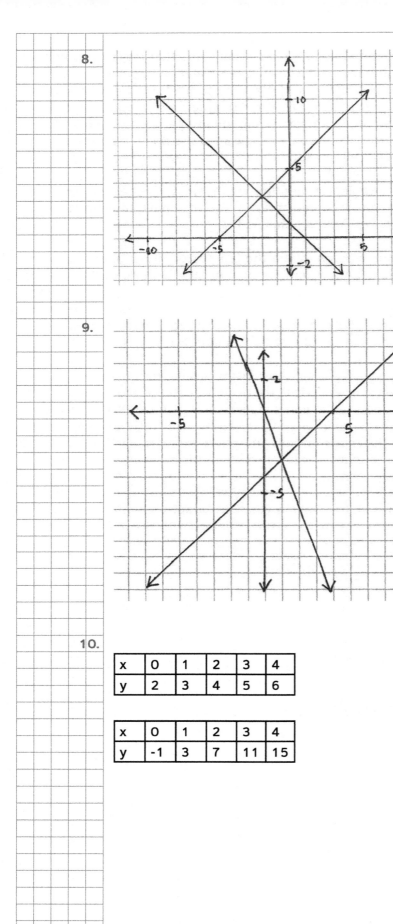

9.

10.

x	0	1	2	3	4
y	2	3	4	5	6

x	0	1	2	3	4
y	-1	3	7	11	15

11.

x	0	1	2	3	4
y	4	6	8	10	12

x	0	1	2	3	4
y	7	8	9	10	11

12. $y = 3x + 2$
$y = -x - 6$

13. $y = x + 2$
$y = -x + 6$

14. $y = \frac{1}{2}x + 2$
$y = \frac{3}{4}x + 1$

15. $y = 3x$
$y = x - 4$

16. $y = x - 6$
$y = \frac{1}{2}x$

17. $y = 2x - 3$
$y = -4x - 3$

18. Did you really need to graph Problem 17? What feature of the two equations allows their shared solution to be more easily found?

19. $-2x + 3y = 12$
$2x + 3y = 12$

20. $4x - 2y = -12$
$4x + 2y = -12$

21. Write a *Note to Self* that explains *how to solve simultaneous equations by graphing*. As always, the most important thing to do is probably to include at least one clearly illustrated example.

REVIEW

No calculators!

1. 23.72 - 9.83

2. 325.57 + 65.7

3. $54.9 \cdot 1.02$

4. $32.67 \div 2.7$

5. $3.42 \cdot 10^2$

6. $\frac{1}{3} + \frac{7}{8}$

7. $\frac{2}{3} - \frac{5}{8}$

8. $\frac{3}{8} \div \frac{5}{8}$

9. $\frac{1}{3} \cdot \frac{1}{8} \cdot \frac{7}{8}$

10. $7\frac{1}{3} \div \frac{44}{9}$

11. $4\frac{2}{5} \cdot 2\frac{1}{7} \cdot \frac{14}{33}$

12. Write 0.23 as a percent.

13. Write 45.7% as a decimal.

14. What is 12% of 60?

15. 5 is what percent of 30?

16. What is the percent increase between 12 and 20?

17. 3 meters = _____ cm

18. 200 mm = _____ cm

19. 8 liters = _____ ml

20. 10 km = _____ m

21. 4.2 kg = _____ g

22. There was a math exam consisting of two questions only. Fifty percent of the students solved Question 1; 80% of the students solved Question 2. Every student solved at least one question and six solved both. How many students took the exam?

23. Lenny's bus route is exactly one hour long. Every 15 minutes two buses begin the route, one from each end. How many buses will Lenny's bus meet on its entire route?

24. Fourteen is 20% of a number. That number is 25% of a bigger number. Find both of the unknown numbers.

25. Use a variable to solve this problem. (Hint: Setting up a four-by-four table would be a good idea.)

Every month, Colonel Aureliano Buendía and Melquíades each acquire one block of ice. Right now they own a total of 24 blocks of ice. In six months, Melquíades will have twice as many blocks as the Colonel. How many does each man have now?

2 SOLVING BY ELIMINATION

To start this lesson, I'm going to show you a way that you could solve the lizard problem from the last lesson using pictures. (Who knows, maybe you solved it this way on your own.) The illustrations can be tricky to understand, but I really think it's worth taking time to figure out how they work.

The fact that the sum of the two lizards' lengths is **56 centimeters** could be illustrated like this:

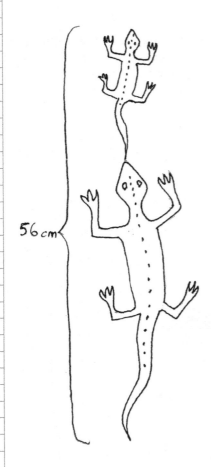

This illustration, of course, goes along with the equation **y + x = 56**.

You could illustrate the fact that the difference between the two lizards' lengths is **22 centimeters** (**y - x = 22**) like this:

22 cm

1. **Explain why this picture illustrates the equation y - x = 22.**

2. Study the combined illustration below closely. It merges the two illustrations that you just looked at. There are two expressions that you could write for the total length of the picture. One of these expressions is suggested by the labels on the left-hand side. How long is the whole illustration, measured in centimeters? The other expression is suggested by the pictures on the right-hand side. In other words, how long is the whole illustration *measured in Florences*? (Remember, Florence is the long one and we've been referring to Florence's length with the variable y.) The two expressions that you just wrote — one with numbers and one with y's — are both expressions for the length of the same picture, so they must be equal to each other, right? Use them to write an equation.

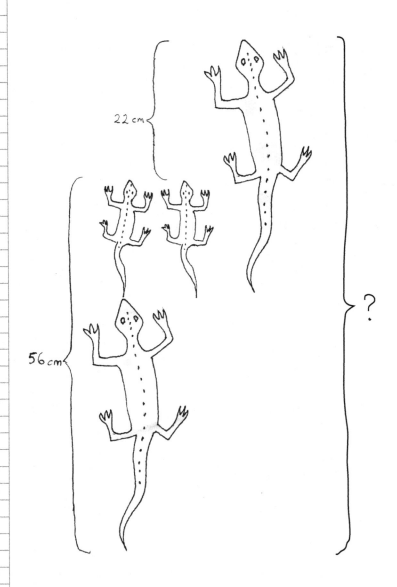

3. Your new equation should have only one variable in it. In fact, it is an extremely easy one to solve, and since y stands for Florence's length, you've essentially just solved the problem without guessing and checking. How long is Florence?

Chapter 3, Lesson 2

4. Go back to one of the two original equations ($y + x = 56$ or $y - x = 22$), plug in your value for y, and use it to find a value for x. How long is Edna?

5. Just to make sure that you have the right values, check that the values you found for x and y make the other original equation true.

You used the illustrations in order to take the two original equations and make a new equation that was also true and that had only one variable and so was easy to solve. You somehow combined $y + x = 56$ and $y - x = 22$ to get $y + y = 78$. (Or you might have written it as $y + y = 56 + 22$ or $2y = 78$ or some other equivalent form.)

Now, I'm not suggesting that you draw one of those illustrations every time you encounter a pair of simultaneous equations. That would take a long time. But there is a way, using the algebra skills you already have, to combine the two equations without using illustrations.

I'm about to claim that you can *add the two equations to each other* and get a new equation that is also true, but I want you to believe that this method really should work, so I'm going to ask you to do something else first.

6. Add the following equations to each other. What I mean is that you should add the elements vertically. For instance, in the first case, add the 6 and the 3, the 13 and the 5, and the 19 and the 8, so that you end up with a third equation that also looks like an addition problem.

$6 + 13 = 19$
$3 + 5 = 8$

Are the original two equations true? Is the third equation that you got also true?

7. Add the following equations in the same way. Be careful about what it means to add two negatives!

$12 - 3 = 9$
$6 - 3 = 3$

Are the original equations true? How about the new equation?

8. Try one more pair. This time remember what happens when you add something to its opposite.

$7 - 2 = 5$
$9 + 2 = 11$

Are all three equations true?

In precisely the same way, you can add two equations with variables in them and get a third equation that's also true. Generally this is only really handy when you can do something similar to what just happened in Problem 8, where one of the pieces disappeared when you added the equations. That's what happens when you add the two lizard equations:

$y + x = 56$
$y - x = 22$

9. Add those two equations together, being careful about positives and negatives. I hope the equation that you got was the same as (or an equivalent form of) the equation that you got when you used the illustrations.

Find the shared solution (point of intersection) for the sets of two equations below by adding them together. I'm going to start you off with two equations from the last lesson. Since you are solving two equations that each have an x and a y in them, *you haven't really solved them until you find both x and y*. You didn't have the solution to the Florence and Edna problem until you found *both* of their lengths. So every time you find the value of one variable, plug that value into one of the equations to find the other variable's value. Then check that your values for x and y make both equations true.

10. $-2x + 3y = 12$
    $2x + 3y = 12$

11. $4x - 2y = -12$
    $4x + 2y = -12$

12. Look back at Problems 19 and 20 from the last lesson. Do your solutions from adding the equations match those from graphing? If so, good job! If not, check your work to see where the error is.

For the following, find the shared solution (remember, the solution means both x and y!) and check your answer to make sure that it satisfies both equations. Notice that I've written "1y" instead of just "y" because I want to make the addition easier for you. I hope you realize that 1y and y are the same thing! So are x and 1x. In your own work, you can choose whether to write down the coefficient 1. Sometimes I write it and other times I don't.

13. $2x + 3y = -12$
    $-2x + 1y = -4$

14. $1x - 2y = -4$
    $3x + 2y = 12$

15. $4x - 4y = -16$
    $1x + 4y = 1$

16. $1x - 3y = -12$
    $3x + 3y = 0$

17. $-5x - 4y = -20$
    $5x + 3y = 15$

18. $6x + 2y = -24$
    $1x - 2y = 3$

You may encounter simultaneous equations that don't line up as neatly as the ones you've had so far when you arrange them vertically. In such cases, you'll have to rearrange one or both of the equations into a new form so that the **x-elements**, the **y-elements**, and the constants line up properly.

Rearrange the following equations using the techniques that you know, then add them to find their shared solution.

19. y = x + 2
    x + 2y = -2

20. 2x + 3y = 11
    y = 2x - 7

21. 2y = -2x + 16
    7x - 2y = 20

22. -4x + 5y = 22
    3x = 5y + 1

In this next problem, one of the variables has a fractional value. Don't freak out!

23. y = 2x + 4
    2x = -5y +14

24. The form of the equations that I gave you in Problems 13 through 18 (for example, 3x + 2y = 15) is called *standard form*. You may well have converted Problems 19 through 23 into that same form. At any rate, you should know and remember the name of that form, so take a minute to make a *Note to Self* that explains it and gives a couple of examples. The generic version of standard form is this: ax + by = c. (Here, the a, b, and c are meant to stand for constants — regular old numbers — and the x and y for variables.)

How about if I were to ask you to solve two simultaneous equations such as **x + y = 27** and **2x + y = 39**?

25. **Add the equations x + y = 27 and 2x + y = 39 in the same way that you've been doing.**

The difficulty here, of course, is that the new equation still has two variables in it. It's no easier to solve than either of the original two equations. The way out of this difficulty is that, rather than *adding* the two equations, you can *subtract* one from the other.

26. **Subtract one of the two equations in Problem 25 from the other. It actually doesn't matter which one you subtract, as long as you subtract every element of the one from the other and continue to be careful about negative numbers. The new equation should have just one variable; use it to solve the problem as you've been doing. If your solution doesn't make both of the original equations true, you've probably made an error in your subtraction and you should try again from the beginning or ask for help.**

The only reason that using subtraction can be a little more difficult than using addition is that you can get tripped up, especially when you have to subtract a negative. But you can deal with it. The great thing about these problems is that you can be absolutely sure your solution is correct by checking it in the original equations.

Solve the following systems of equations by addition or subtraction. Check your answers.

27. $-2x + 1y = 8$
$2x - 8y = 6$

28. $1x + 1y = -4$
$1x + 8y = 3$

29. $-8x + 2y = 48$
$1x + 2y = 3$

30. $y = x + 6$
$x + y = 2$

31. $y = 5x + 10$
$5x + 3y = 10$

32. $-2x + 5y = 15$
$5y = 25x - 100$

33. $-x + y = 0$
$y = 2x - 4$

34. $-4x - 3y = 0$
$y = -4x - 8$

Now consider a pair of equations such as $x + 2y = 1$ and $3x - y = -18$.

35. Try adding those two equations. Then try subtracting them.

36. As you just saw, the difficulty here is that neither adding nor subtracting eliminates one of the variables. Once again, I think you have all of the tools to figure out how to overcome this difficulty yourself. Spend at least ten minutes working on this before you seek help. I'd recommend working with partners. (Here's your hint: You know that you can transform any equation into an equivalent equation as long as you do exactly the same thing to both sides. You'll have to transform one of these equations into an equivalent. Transforming either one will work.)

I bet you figured that out by yourself, but I'll explain it just in case you didn't. The key is to *multiply* one of the equations by something so that when you add or subtract them, the **x** or **y** will be eliminated. You had a number of choices, but here are two.

37. Multiply both sides of the equation $3x - y = -18$ by 2. Be careful to multiply *all three elements* by 2. Take the new equation that you come up with and add it to $x + 2y = 1$. The y-element should disappear and you should be able to solve as usual. Check your solution in both of the original equations!

38. Solve the same pair of equations by multiplying $x + 2y = 1$ by 3 and then subtracting one equation from the other. This time you'll eliminate the x, but you'll get the same solution.

Solve the following systems of equations. Check your work. In some cases you may need to multiply each equation by a different number in order to eliminate a variable.

39. $-4x + y = 1$
$5x + 2y = 15$

40. $-2x + 5y = 5$
$3x - 2y = 9$

41. $1x - 5y = -15$
$y = -2x + 14$

42. $-5x + 3y = 9$
$x - 5y = 7$

43. $y = -\dfrac{3}{4}x + 3$

$6x + 11y = 42$

44. $3x + 4y = 12$

$4x + 3y = 2$

45. $x + y = 6$

$-4x + 5y = -15$

46. $3x + 5y = 19$

$2x - 2y = -14$

47. On a recent backpacking trip to Waldo Lake, Greg's family encountered a mosquito problem: there were clouds of them and they were all hungry! His daughter and her friend were comparing bites at the trailhead. They counted 4 bites on Celia; her friend already had 7. As they hiked on, it seemed that the mosquitoes were taking a greater liking to Celia than to her friend; Celia was bitten at an average rate of 10 times per 3 minutes and her friend was bitten at a rate of 3 times per 1 minute. At this rate, how long before the two of them shared the same number of bites? Use the techniques you have practiced in this lesson to solve this system of equations. Think about each hiker's constant (starting number of bites) and her rate of change. Make y equal to the total number of bites.

All of the techniques that you've practiced in this lesson come under the heading of solving simultaneous equations *by elimination*, since they all involve manipulating the equations so as to eliminate one of the variables.

48. Write a *Note to Self* explaining *how to solve simultaneous equations by elimination*. Make sure to show examples that include multiplying one or both equations as well as adding and subtracting.

REVIEW

1. Matt wants to create a pizza that has all of his favorite toppings in the proper ratios, but doesn't like more than one topping per slice. He likes pepperoni a lot, so he makes half of the pizza with only that; 1/8 will have just olives; 1/3 will have just sausage; and the remaining part will have anchovies. What fraction of the pizza will have anchovies on it? In what ratio does Matt like his toppings (A : O : S : P)? Into how many equal pieces must he cut his pizza so that each piece has only one topping?

2. Enrollment at the local cheesemaking school was 357 students last year. This year the enrollment is 315. By what percent has enrollment decreased? Round your answer to the nearest whole percent. No calculators!

3. A teacher graded a week's worth of quizzes. Each of the students received an A, B, or C. The number of quizzes receiving A's, B's, and C's were in the ratio 5 : 3 : 1, respectively. There were 40 more B's than C's. Find the number of quizzes that received a grade of A and the total number of quizzes graded by the teacher.

4. **"Do-Re-Mi"** *from* The Sound of Music, *1959, by Rodgers and Hammerstein*
Do, a deer, a female deer,
Re, a drop of golden sun,
Mi, a name I call myself,
Fa, a long long way to run,
So*, a needle pulling thread,
La, a note to follow so,
Ti, a drink with jam and bread,
That'll bring us back to Do —
do-re-mi-fa-so-la-ti-do!

*The actual syllable is *sol*. Rodgers and Hammerstein cheated just a little bit.

While playing an A, a guitar string vibrates at a rate of 440 cycles per second. If this is true, complete the table for the other notes on the musical scale. Some of the frequencies will be mixed numbers (whole numbers and fractions). Still no calculators!

Note	Expression	Frequency
La (A)	x	440
Ti (B)	9/8 of x	
Do (C)	5/4 of x	
Re (D)	4/3 of x	
Mi (E)	3/2 of x	
Fa (F)	27/16 of x	
Sol (G)	15/8 of x	
La (A)	2x	

5. The formula for the area of a triangle is $A = \frac{1}{2}bh$, where A is the area, b is the length of the base, and h is the height of the triangle. This formula has three variables, so we'll hold one constant by saying that our triangle has a 10 cm base. What is its area if it's 10 cm in height? (Be careful about units!) How about 15 cm? 25 cm?

How does the area vary with the height of the triangle? If you graphed the changing area (the dependent variable) of a 10 cm-base triangle versus its changing height (the independent variable), what would be the slope of the resulting line?

6. Tweedledee and Tweedledum each lose a rattle every day. Right now Tweedledee has five times as many rattles as Tweedledum. In one week they'll have a total of 52 rattles. How many does each have now?

3 SOLVING BY SUBSTITUTION

In this lesson you'll learn one more method for solving simultaneous equations, called *solving by substitution*. (It's called that for reasons that I hope will become clear.) It's a third tool in your bag, along with graphing and elimination, and in most situations it'll be up to you which one to use.

To use this method you must first believe that an equal sign means that two things are precisely the same and that you can replace one with the other.

Let's consider the simultaneous equations $x + 2y = 2$ and $2x + 3y = 6$. Now, as you know, one option would be to multiply one of the equations by something and then add or subtract them. But let's look at another option.

1. **Take the equation $x + 2y = 2$ and make an equivalent equation by subtracting 2y from each side. (The new equation should be in the form x = _____.)**

Now, if you really believe that the equal sign means that **x** and "_____" are exactly the same, then you should be able to do this:

2. **Replace the x in the equation $2x + 3y = 6$ with the expression that you know x is equal to. Put the expression in parentheses to indicate that the whole expression gets multiplied by 2. Your new equation should be in the form 2(_____) + 3y = 6.**

3. **Your new equation once again contains just one variable. You can solve this. The only thing to be careful of is that you need to apply the Distributive Rule as you multiply 2 by the expression in parentheses. Remember that you also need to be careful about multiplying with negative numbers.**

4. **As always, now that you have a value for y, use one of the original equations to find a value for x and then check that those two values also make the second equation true.**

That's all there is to the substitution method: you simply isolate one of the variables and then substitute the expression that it's equal to into the other equation. This method will work for any sets of simultaneous equations, though for some sets it's easier than for others. Be careful which variable you choose to isolate: if that variable has a coefficient, things may get complicated.

There's one other thing to watch out for when you're solving simultaneous equations. Consider this pair:

$y = 4x - 2$
$3x - 2y = 24$

I hope it's obvious that the easiest route is to replace **y** in the second equation with **(4x - 2)**, since you know that **y** is equal to **(4x - 2)**. I'll take that step for you: $3x - 2(4x - 2) = 24$.

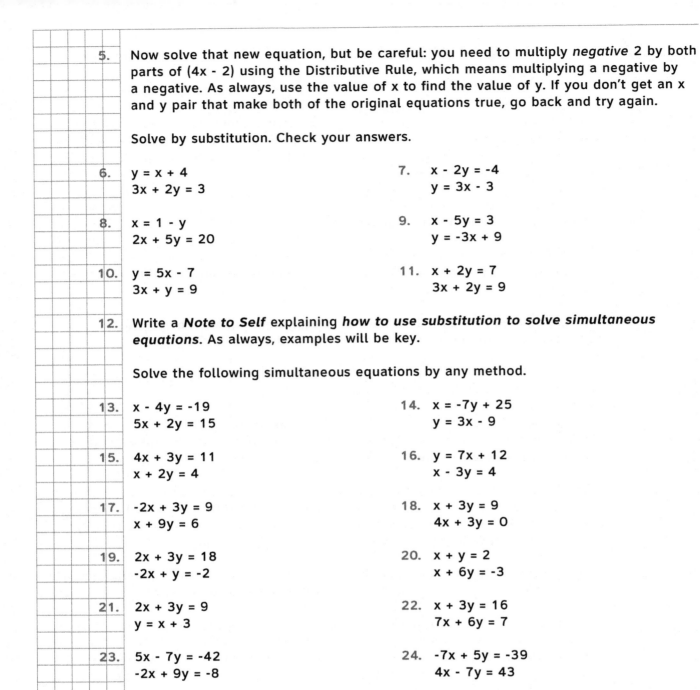

5. Now solve that new equation, but be careful: you need to multiply *negative* 2 by both parts of (4x - 2) using the Distributive Rule, which means multiplying a negative by a negative. As always, use the value of x to find the value of y. If you don't get an x and y pair that make both of the original equations true, go back and try again.

Solve by substitution. Check your answers.

6. y = x + 4
   3x + 2y = 3

7. x - 2y = -4
   y = 3x - 3

8. x = 1 - y
   2x + 5y = 20

9. x - 5y = 3
   y = -3x + 9

10. y = 5x - 7
    3x + y = 9

11. x + 2y = 7
    3x + 2y = 9

12. Write a *Note to Self* explaining *how to use substitution to solve simultaneous equations.* As always, examples will be key.

Solve the following simultaneous equations by any method.

13. x - 4y = -19
    5x + 2y = 15

14. x = -7y + 25
    y = 3x - 9

15. 4x + 3y = 11
    x + 2y = 4

16. y = 7x + 12
    x - 3y = 4

17. -2x + 3y = 9
    x + 9y = 6

18. x + 3y = 9
    4x + 3y = 0

19. 2x + 3y = 18
    -2x + y = -2

20. x + y = 2
    x + 6y = -3

21. 2x + 3y = 9
    y = x + 3

22. x + 3y = 16
    7x + 6y = 7

23. 5x - 7y = -42
    -2x + 9y = -8

24. -7x + 5y = -39
    4x - 7y = 43

25. -8x + 3y = 18
    y = x - 4

26. 11x + 7y = 42
    8x + 5y = 32

So far I've generally given you simultaneous equations where **x** and **y** are both integers. However, Zeus did not descend from his throne on Mount Olympus and decree that all sets of simultaneous equations should have whole-number answers.

Solve the following sets of simultaneous equations using whatever methods you prefer. You'll get fractional answers; check them in the original equations just as you have been doing.

27. $2x + 3y = 2$
$12x = 6y + 4$

28. $x = 6y$
$2x + 3y = 3$

29. $x + 2y = 1$
$y + \frac{1}{6} = \frac{1}{2}x$

30. Create four sets of simultaneous equations. Make sure that they share whole number solutions. Give them to your classmates to solve.

31. Here's a challenge for you. I'd recommend doing this problem with your math partners. You now have all of the necessary equipment to tackle a set of three simultaneous equations, each of which has three variables. You can use any combination of the techniques you've learned. (Actually, graphing would be very hard, since your graphs would need to be three-dimensional, but some combination of elimination and substitution should work well.) Be flexible in your thinking, try hard, and work on this for at least twenty minutes before you give up. After twenty minutes, or after you've pulled out more than one handful of hair, seek help from a teacher. Here are the three equations:

$x - 2z = -3y - 7$
$4x - y = 2z + 16$
$10 + 3z = 3x - 2y$

Algebra has been developed over the course of many centuries. In that time, lots of different ways of expressing equations have been tried, and some of them are useful in various situations. Here is a way of writing linear equations called *parametric form* that you won't study intensively until a much more advanced math class:

$$x = (-2, 1) + \lambda(5, 5)$$
$$x = (1, 4) + \lambda(3, -3)$$

The graphs of those equations are both straight lines. Here's one of them:

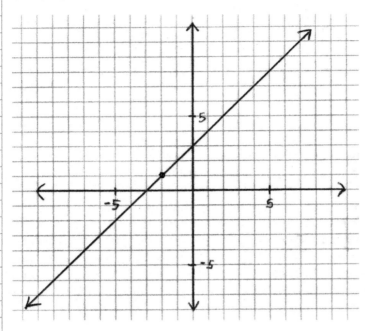

32. Which equation do you think that graph shows? Why? What does the λ sign indicate? Based on what you've concluded, graph the other line and find their shared solution.

REVIEW

1. Find the Least Common Multiple (LCM) of the following numbers.
   9 and 27   8 and 10

   Do the following division problems. Check your answer using multiplication.

2. $1.44 \div 0.09$ 3. $11.2 \div 0.07$

4. Fifteen girls and five boys each write their name on a slip of paper and place it in a hat. What is the chance of selecting a girl's name from the hat? Express your answer as a common fraction.

5. Convert the following measurements.

45 L = _____ mL

2.2 km = _____ m

321 cm = _____ m

6. In how many different ways can 30 cents be made by combining quarters, dimes, nickels, and/or pennies?

What number will make each fraction equivalent?

7. $\dfrac{4}{15} = \dfrac{n}{75}$

8. $\dfrac{m}{6} = \dfrac{12}{36}$

9. $\dfrac{5}{x} = \dfrac{35}{14}$

Find each product. Check your answer using division.

10. $(0.7)(210)$

11. $(3.2)(1.4)$

12. There are 6 euros to 1,200 Japanese yen. How many euros could you exchange for 3,000 yen?

13. What is the area of a 12-meter by 11-meter dance floor?

14. Six knights gathered for a jousting tournament. Work out the ranking of the knights, the color of each man's horse and lance, and the Order he represents. (Warning: This is a hard puzzle. If you find yourself spending a crazy amount of time on it, go ahead and move on. It's okay if these aren't your cup of tea.)

1. Sir Palamon did better than Charles the Bald.
2. The knight who rides a gray horse carries a purple lance.
3. Charles the Bald placed two spots below Don Quixote, who was not as good as the knight on the chestnut horse.
4. The knight who rides a white horse finished just above the knight who carries a green lance.
5. The knight with the roan horse finished last.
6. The Black Prince finished higher than the knight from the Order of the Barking Deer but lower than the knight with the purple lance.
7. The knight from the Order of the White Bear rides a chestnut horse.
8. The knight from the Order of the Chafing Garter placed third, which was better than the knight with the striped lance.
9. The knight on the white horse finished two spots below the knight from the Order of the Silver Parrot.
10. The knight on the black horse (who is not The Black Prince) finished second.
11. Sir Roland carries a blue lance.
12. The knight on the bay horse finished above the knight from the Order of the Armored Codpiece but below the knight with the red lance.
13. The knight with the red lance was not the champion.
14. Sir Bedevere finished two spots below the knight from the Order of the Golden Fleece.
15. The knight from the Order of the White Bear was better than the knight on the gray horse, who was better than the knight with the yellow lance.
16. The knight with the yellow lance finished behind Charles the Bald.

4 SHARED SOLUTIONS

1. I'd like to start this lesson off by asking you a question: Do all pairs of two-variable equations have a shared solution? The best way to prove that a statement is false, of course, is to find a single counter-example. If you can find just two equations that *don't* share a solution, then not *all* pairs of equations have a shared solution.

   Spend at least ten minutes testing some equation pairs. One way to go about it would be to pick pairs more or less at random and test them. If you do this, be sure not to pick the pairs that have been used in the last three lessons — as you well know, those pairs definitely *do* have shared solutions. For me, the best way to consider this question is to think about the *graphs* of equations and what those graphs show about the equations' solutions.

   I imagine that you were able to figure out the answer to that question, but once again I'll take you through a possible solution. Consider this pair of simultaneous equations: 2y - 4x = -8 and y - 2 = 2x.

2. In my experience, most students don't tend to pick graphing as their favorite method of solving simultaneous equations — it's a little too laborious. Nevertheless, I'm going to ask that you graph these two equations according to the method that you learned in Lesson 1 of this chapter. What do you observe about the lines that represent these two equations?

   Parallel lines, as they're defined mathematically, go on forever without touching. If each of the lines you drew represents the solutions to one of the equations, I hope you can see that they do not share any solutions. When two equations do not share any solutions, they are called an *inconsistent* pair of equations. (Based on this, I'm sure you won't be surprised to hear that if two equations *do* share a single solution, they're called *consistent*.)

3. Of course, as I said, you probably don't usually use graphing to solve simultaneous equations. So how can you tell that a pair of equations is inconsistent when you are solving by elimination or substitution? Well, try it. You know that the pair 2y - 4x = -8 and y - 2 = 2x are inconsistent, so try solving them using both elimination and substitution and see what happens.

   As you just saw, when you try to solve inconsistent equations using elimination or substitution, you end up with a statement that is simply untrue, such as 4 = 8 or 0 = -12.

   So, a pair of equations can be consistent or inconsistent. There is a third possibility as well. Consider the pair y - 1 = 3x and 3y - 9x = 3.

4. This time, solve that pair by elimination or substitution first.

   Once again, you ended up eliminating all of the variables in the equations. But this time, instead of ending up with an equation that is false, you ended up with one that is true.

5. Imagine graphing those two equations. What do you think you'll discover about their graphs?

6. Test your hypothesis by actually graphing the two equations.

As you saw, it is possible for two equations to be represented by *exactly the same line*, in which case they share an infinite number of solutions. In fact, when this happens, the two equations are really equivalent versions of the same equation and so, as a pair, they are called *equivalent*.

7. Write a **Note to Self** explaining about *inconsistent, consistent, and equivalent simultaneous equations*. Explain how you can recognize the three different varieties from their graphs and when you are solving them by elimination or substitution.

Carefully examine the following pairs of linear equations. What method — graphing; substitution; or elimination by addition, subtraction, multiplication, or some combination of operations — would be most efficient to solve each system? You do not need to solve these; just explain which method you would choose and why.

8. $x + 5y = 18$
   $-2x + 5y = 9$

9. $y = x - 2$
   $2x + y = 7$

10. $3x - 2y = 4$
    $x = 2$

11. $-2x + 5y = 4$
    $2x + 3y = 12$

12. $7x + 2y = 17$
    $2x + 3y = 0$

13. $-7x + y = -2$
    $y = -x - 2$

14. $-5x + y = 12$
    $y = 5x + 12$

15. $5x - 4y = -3$
    $x + 6y = 13$

16. $y = 3x + 4$
    $y = x + 4$

Solve the following pairs of simultaneous equations using whatever method you like. Be prepared: Some pairs are inconsistent or equivalent.

17. $6x + 5y = -22$
    $4x + 3y = -16$

18. $2x + y = -8$
    $y = -2x - 8$

19. $x + y = -4$
    $-2x + y = 5$

20. $x + y = 7$
    $2x + 2y = 10$

21. $2x + 3y = 16$
    $4x + y = 2$

22. $x + 3y = 14$
    $y = x + 2$

23. $y = \frac{1}{2}x + 1$
    $x - 2y = -5$

24. $13x + 7y = -120$
    $4x + 5y = -17$

25. $11x - 13y = 12$
$-2x + 15y = 61$

26. $x = -2\frac{1}{2}y + 17.5$
$2x + 5y = 35$

27. $2x + 5y = 40$
$4x = 40 - 10y$

28. $5x - 4y = -6$
$-3x + 7y = 45$

29. $7x - 3y = -24$
$-5x + 8y = 64$

Solve the following questions by writing and solving simultaneous equations.

30. The sum of two numbers is 5 and their difference is 11. What is the product of the two numbers?

31. The sum of Jim's weight and Bob's weight is 180 pounds. If you add Jim's weight to Bob's weight, you get a number that is three times Bob's weight. How many pounds does Bob weigh?

32. There are only bicycles and tricycles in Greg's backyard. He counted a total of 30 seats and 70 wheels. How many tricycles are in the backyard?

33. Scott wants to build stairs from his second-story deck to the yard below — well, *he* doesn't want to build stairs, he just wants them built. He calls up a couple of construction companies to give him bids (estimates of how much they'll charge) for the work. Kwality Konstruction would charge $360 plus $12 per hour. Better Buildings charges $280 plus $15 per hour.

   a. Write an equation for each bid, letting x represent the number of hours it would take to build stairs and letting y represent the total cost.
   b. What amount of construction time would cost Scott the same amount of money whichever company he chooses, and what would that cost be?
   c. Providing both companies do equally good work and both predict that the job will take 20 hours to complete, which company should Scott hire and how much would he save?

34. Do you think that there are more possible pairs of consistent equations in the universe than pairs of inconsistent or equivalent equations? Write about your ideas and be prepared to discuss them with your classmates.

35. It's possible, as you saw in the last lesson, for three equations with three variables apiece to share a single solution. Is it possible for three equations with *two* variables apiece to share a single solution? If it is possible, sketch a graph of what such a situation might look like. Does it seem *likely* that such a situation would occur? Why or why not?

36. In this chapter you've spent a lot of time looking at how two straight lines can interact. Every time you've solved a pair of simultaneous equations you've been dealing with two straight lines, even if you haven't graphed them. As you know, two straight lines can cross (making them consistent), run parallel to each other (inconsistent), or be on top of each other (equivalent). You know that not all equations have straight line graphs. So, if you had a pair of simultaneous equations where one was a straight line but the other wasn't, how many shared solutions could they have? Here's a sketch of the graph of $y = x^2$. Copy it into your notebook:

Now try putting straight lines on that graph to see how many solutions they could share with $y = x^2$. You should find three different *numbers* of solutions that could be shared (in the same way that two straight lines can share one solution, no solutions, or infinite solutions). Make a clear illustration for each of these possible situations.

REVIEW

1. If the length of the edge of a cube increases by 50%, what is the percent increase in the volume of the cube? Express your answer to the nearest whole percent. (It may be helpful to use real numbers — if a cube has an edge of two inches, what's its volume? What happens when that two-inch edge increases by 50%?)

2. Three angles of a triangle have a ratio of 1 : 3 : 14. How many degrees is each angle?

Apply the Distributive Rule to the following:

3. $6(x - 5)$

4. $y(x + 10)$

5. $3x + 6$

Solve the following equations:

6. $2x + 8 = 3x + 7$

7. $3x = 4 - 3x$

8. Herr Settembrini's class gains two students a week; Herr Naphta's gains one student a week. Right now Herr Naphta has three times as many students as Herr Settembrini; in three weeks he'll have seven more students than Herr Settembrini. How many students does each pedagogue have?

9. Every day Madame Chauchat gives Hans Castorp a pencil. Today they have the same number of pencils, but in six days he'll have twice as many pencils as she does. What's the total number of pencils owned by the pair?

10. The formula for the number of diagonals in a polygon is $D = \dfrac{n(n-3)}{2}$, where D is the number of diagonals and n is the number of sides.

How many diagonals does a six-sided polygon have? How about a ten-sided polygon? Twenty-sided? How many does a three-sided polygon have? What happens when you plug two into the formula? Why does this result make sense, given what you know (or what you can figure out — try drawing one) about two-sided polygons?

You're familiar with three types of variation. How does the number of diagonals in a polygon vary with the number of sides? (This is actually a pretty tricky question. It's probably not obvious at first glance which of the three families this equation falls into. You might consider making a chart of the values of D for the first few values of n greater than three.)

11. A problem from Harry Edwin Eiss's *Dictionary of Mathematical Games, Puzzles, and Amusements*:

You have eight pennies in a row, like so:

Your job is to use four moves and end up with four stacks of two pennies each. Every time you move a penny, it must skip over two pennies and land on top of the next penny after that. For instance, you could place Penny 4 on top of Penny 7 or on top of Penny 1. Skipping over a stack of two pennies does count as skipping over two pennies: if you had already stacked a penny on top of Penny 5, you could place Penny 4 on top of Penny 6.

5 MIXTURE PROBLEMS

In the very first lesson of Chapter 1, I gave you a problem to solve by guess-and-check and told you that I'd come back to show you how to use algebra to solve that problem later on. Well, "later on" is now, because that problem is a classic *mixture problem* and the algebraic way to solve it is to use simultaneous equations. Here it is again:

I have 29 coins (all nickels and dimes) totaling $2.15. How many nickels do I have?

(Actually, you've solved at least a couple of other mixture problems in this book: a zoo with one-eyed men and three-headed dogs comes to mind. And remember all those bikes and trikes in Greg's yard from the last lesson?)

There are only two tricky things about solving a mixture problem: first, recognizing that you are dealing with a mixture problem, and second, writing the two equations that go with the problem. After that, you've simply got two simultaneous equations, and you've been practicing methods for solving two simultaneous equations throughout this chapter.

So let's take a look at the coins problem. It's a mixture problem because you're mixing two different things together (nickels and dimes) and because you have information that tells you about the quantity and the value of those things. What you need to do is write two equations that describe those quantities and values. For this problem, we'll use **x** to stand for the number of nickels and **y** to stand for the number of dimes. The equation that describes the quantity (or "amount" or "number") of coins is easy. There are **29** total coins, **x** of which are nickels and **y** of which are dimes, so... **x + y = 29**.

1. **The equation for the value of the coins is slightly harder to write. You actually did it back in the first lesson of Chapter 1, so you could just look back through your notebook, but I think it's useful to do that work again now that you have some more experience. So, write the equation for the value of the coins. Here's what you should keep in mind: the value of each nickel is 5 cents, the value of each dime is 10 cents, and the total value of the coins is 215 cents.**

You now have two equations that describe the coin situation. If you only had one equation, say, **x + y = 29**, there would be an infinite number of possible answers. (Actually, the number of answers would be infinite only if there were such things as negative coins or fractions of coins. But even with only positive, whole number coins, there are a lot of possible answers. You could have **28** nickels and one dime, **27** nickels and two dimes, and so on.) However, you just wrote an equation that is true at the same time that **x + y = 29** is true, and there's only one solution that those two equations share.

2. **Solve those two simultaneous equations, using whichever method you like best. Check to see that your solution is correct: do you have 29 coins and is their total value $2.15?**

Here's another one:

3. You're making a magic substance consisting of a blend of fairy dust and moose extract. You can get fairy dust at the local market for 16 copper coins per gram. Moose extract will run you 50 copper coins per gram. (That stuff is expensive. Do you realize how hard it is to harvest?) Anyhow, the magician that you're apprenticing with gives you 1,762 copper coins (in a very large purse), which he tells you will purchase ingredients for exactly 40 grams of the final magical substance. How much fairy dust do you need to buy and how much moose extract?

The first thing to do is to recognize that this is a mixture problem. It is, of course — not only because it's in a lesson called "Mixture Problems" but because you're mixing two things and you know about the total quantity (40 grams) and the total value (1,762 copper coins). So write your two equations. As before, the quantity equation should be pretty straightforward, and the value equation should be a little trickier. Once you've written your two equations, they're simultaneous, so go ahead and solve them. Check that your final product really would weigh 40 grams and cost 1,762 copper coins.

Good work. Here are a few more mixture problems:

4. Your magician master gives you a crate containing both three-winged and five-winged mice. There are 28 animals and 98 wings. How many of each variety of mouse has he given you? (The only trick here is that the "value" of the animals is not measured in money, but in wings.)

5. A gameshow contestant gets 2 points for the correct answer to an easy question and 5 points for the correct answer to a hard question. If he answers 20 questions correctly for a total score of 61, how many of each type of question did he get right?

6. A certain made-up metal called woodchuckium is an alloy of two other types of made-up metals, weaselium and superbium. (If you don't know what an alloy is, look it up.) Ten cubic centimeters of woodchuckium weigh 57 grams. One cubic centimeter of weaselium weighs 7.9 grams and one cubic centimeter of superbium weighs 3.5 grams. How many cubic centimeters of each of the two other metals do you need to make 10 cubic centimeters of woodchuckium? (Hint: One equation will represent the *volume* of the total; the other will represent the *mass* of the total.)

7. On a recent trip to Bhutan, a friend of mine spent 1,284 ngultrum on two different varieties of plant. (There is plenty of fiction in this book, but Bhutan is a real country and its currency really is the ngultrum. My friend is made up, though.) One kind of plant cost 11 ngultrum apiece and the other cost 15 ngultrum apiece. If she bought 100 plants, how many of each variety did she get?

8. In my pocket, I have 14 coins of two types: Malaysian ringitts and Estonian kroons. (Once again, real places and real kinds of money.) One ringitt is worth 31 cents and one kroon is worth 8 cents. (That's pretty close to accurate on the day I'm writing this problem as well.) If the total value, in dollars, of what's in my pocket is $2.96, how many of each kind of coin do I have?

9. You're filling the wading pool in your backyard with Jello. Lime Jello costs 75 cents a pint and cherry Jello costs 45 cents a pint. It takes 3,000 pints to fill the pool and you spent $1,762.50 on Jello. How much of each flavor did you use? What color do you think the resulting mixture will be? What are your parents going to do to you when they find out you've got hold of one of their credit cards and are using it for this purpose?

10. Write your own mixture problem for your classmates to solve. It doesn't necessarily need to have whole-number answers, but let's say that at the most the answers can go to two decimal places.

11. Write a **Note to Self** on **how to solve mixture problems**. Include an example, which can be one of the ones I wrote or the one you wrote.

REVIEW

Calculate and round to the nearest tenth:

1. $18\overline{)7}$

2. $3.3\overline{)16}$

Solve the following equations and inequalities for y:

3. $2y - 8x + 3 = 10x - y - 3$

4. $2y - 2x^2 = 16x - 8$
(Don't be fooled into thinking that you can add or subtract an x^2 from an x!)

5. $1 - \dfrac{y}{2} \geq 2x$

6. $10x - 2y < 3y - 5x + 10$

Graph the following equations and inequalities:

7. $3x + y = 2x + 5$

8. $4 - 2y > -6x$

9. $-5y - x \leq -5$

10. Two armadillos are touring the Scottish coast on motor scooters. One of them travels 5 meters per second faster than the other. Several bored sheep clock the armadillos riding for 8 seconds and calculate that they have traveled a combined distance of 184 meters. How fast is each armadillo going? Remember that a chart may help:

	Armadillo 1	Armadillo 2
distance		
rate		
time		

11. The table below shows the average life expectancy of denominations of American currency. In other words, it's how long each kind of bill tends to last before it falls apart:

Value of Currency ($)	1	5	10	20	50	100
Time in Years	1.5	1.25	1.5	2	5	8.5

Make a graph of this data. For the vertical axis, I'd suggest a scale of two graph paper squares to one year. The horizontal axis is a little funny: you should just go ahead and place each denomination in order of increasing value, each two squares apart. Which of the three types of variation that you've studied does this graph appear to fall into? Based on this graph, roughly how long would you expect a $500 bill (the next denomination up from $100) to last? What explanation would you give for the overall shape of the graph? That is, why is the $5 bill the lowest point and why do the life spans increase for the higher bills? Why is the rate of increase a curve?

There is such a thing as a $2 bill, although they're quite rare. Would you expect the life span of the $2 bill to fit neatly into this graph between the $1 and the $5, or would it break up the lovely curve? Why? Did you know that Grover Cleveland is on the $1,000 bill, that the largest bill ever printed in the U.S. was the $100,000 bill, and that private citizens are not legally allowed to own $100,000 bills? (Okay, you don't have to answer the "did you know" questions.)

12. Another challenge based on one by Harry Eiss. There are three moose-herding collies positioned at points around the moose corral shown below. Each dog runs a loop around four moose and returns to her starting point. Can you trace the three dogs' paths so that every moose is separated from every other moose?

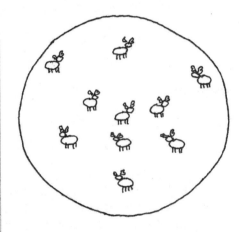

6 SIMULTANEOUS INEQUALITIES

Just as you can have simultaneous equations, you can also have simultaneous inequalities. The key difference is that you cannot solve a set of simultaneous inequalities using either the elimination or substitution methods. Let's look at why this is the case.

1. **Consider the inequalities $y > x - 3$ and $y < 2x + 5$. Try solving them by elimination. What problem do you run into? Why can't you solve them by substitution?**

Think about the equation $y > x$. If you remember, we graph with a dashed line because y cannot be equal to x. We shade above the line to show that y is greater than x.

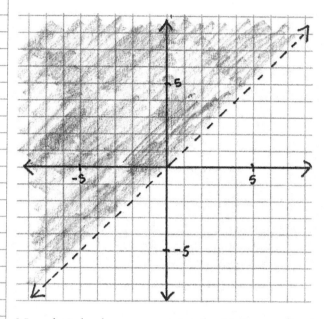

Now let's look at $y > -x$. Again this inequality is "greater than," so it should be a dashed line with shading above the line.

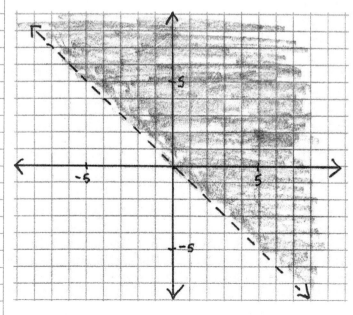

In each case, the graph represents all of the solutions that make each of the inequalities true. Now, if **y > x** and **y > -x** are *simultaneous* inequalities, the question is, "Which solutions make both equations true?"

2. **Study the graphs of y > x and y > -x. Make a graph that you think represents the shared solutions of these two inequalities.**

In a sense, what you end up with is a Venn diagram of the two inequalities. Venn diagrams are those overlapping circles that you've probably seen or even used in some problem-solving situations. Here's a classic Venn diagram that represents a group of weasels, some of which wear hats, some of which wear coats, and some of which wear both coats and hats:

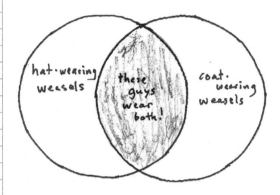

(By the way, Venn diagrams got their name from their creator, a man named John Venn.)

Graphs of simultaneous inequalities work a lot like that: the segment of the graph where the two regions overlap is where you'll find all the solutions that satisfy both inequalities. So I hope your answer to Problem 2 looked something like this:

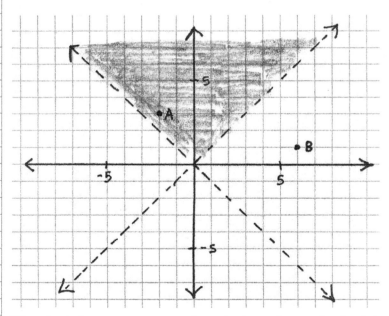

The shaded area represents the solutions that make *both* inequalities true. Just to see whether this really works, look at point **A(-2, 3)**. It lies in the area where the two sets of solutions overlap, so it should make both inequalities true.

Chapter 3, Lesson 6

3. Check to see that (-2, 3) really does make y > x and y > -x true.

4. Point B(6, 1), on the other hand, should make only *one* inequality true — not both. Check to see that this is the case.

Graph the following systems of inequalities.

5. $y > x + 3$
$y < -2x + 9$

6. $y < 4x - 3$
$y < -1x + 7$

7. $y \leq -4x + 5$
$y \geq -1x + 5$

8. $y \geq x$
$y < -3x - 8$

9. $y > 2x + 3$
$y < -1x - 8$

10. $y > -2x + 1$
$x < 1$

11. $y < 5x - 6$
$y \geq -2x + 1$

12. $y < -3x - 4$
$y < x + 4$

13. $y > x + 3$
$y \leq x - 2$ (Be careful with this one! Shade only the solutions that they share!)

Write the set of inequalities that goes with each illustration:

14.

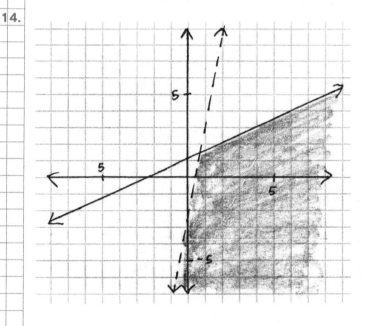

Chapter 3, Lesson 6

15.

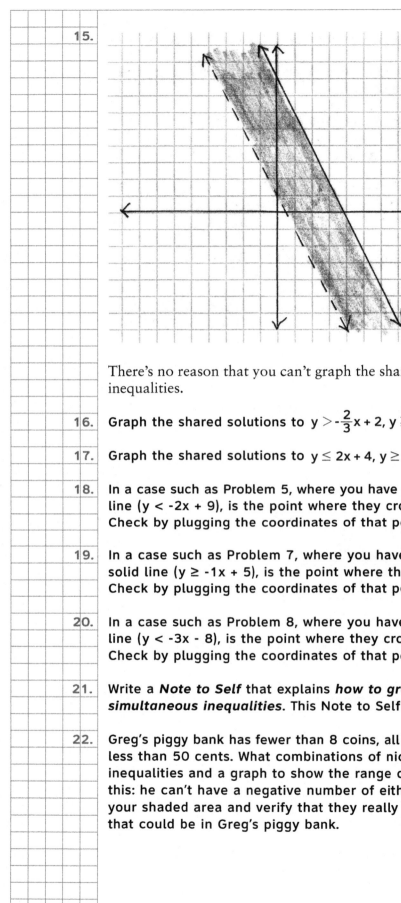

There's no reason that you can't graph the shared solutions to three (or even more) inequalities.

16. Graph the shared solutions to $y > -\frac{2}{3}x + 2$, $y \geq x - 2$, and $y \leq \frac{1}{3}x + 4$.

17. Graph the shared solutions to $y \leq 2x + 4$, $y \geq 2x - 4$, and $y \geq 0$.

18. In a case such as Problem 5, where you have a dotted line ($y > x + 3$) crossing a dotted line ($y < -2x + 9$), is the point where they cross included in the set of solutions? Check by plugging the coordinates of that point into both inequalities.

19. In a case such as Problem 7, where you have a solid line ($y \leq -4x + 5$) crossing a solid line ($y \geq -1x + 5$), is the point where they cross included in the set of solutions? Check by plugging the coordinates of that point into both inequalities.

20. In a case such as Problem 8, where you have a solid line ($y \geq x$) crossing a dotted line ($y < -3x - 8$), is the point where they cross included in the set of solutions? Check by plugging the coordinates of that point into both inequalities.

21. Write a **Note to Self** that explains *how to graph the shared solutions of a pair of simultaneous inequalities*. This Note to Self will need to involve some illustrations.

22. Greg's piggy bank has fewer than 8 coins, all of which are nickels or dimes, totaling less than 50 cents. What combinations of nickels and dimes might be in there? Use inequalities and a graph to show the range of possible solutions. Be careful about this: he can't have a negative number of either coin. Pick two points that lie inside your shaded area and verify that they really are combinations of dimes and nickels that could be in Greg's piggy bank.

23. Given

$x \geq 4$

$y \geq 3$

$x + y \leq 9$

and $y - x \geq -2$

... graph the region that will satisfy all four inequalities.

REVIEW

Solve the following pairs of simultaneous equations:

1. $y = 2x - 8$
$3y + x = -3$

2. $10y - 2x = 20$
$3y - 3x = -12$

3. $4y = -x + 24$
$3x + 12y = 72$

4. Each week Ahab collects two kilograms of ambergris and Queequeg collects three kilograms. Right now Ahab has five times as many kilograms as Queequeg does, but in 28 weeks they'll have the same amount. How much ambergris does each man have?

5. Queequeg can row at a speed of 15 nautical miles per hour; Ahab can only row 10 nautical miles per hour. (His one-leggedness slows him down.) If Ahab gets a three-hour head start and they both row in the same direction, how far will they have traveled before Queequeg catches up?

6. When Greg goes to visit his Canadian relatives, they talk about the temperature in terms of degrees Celsius. Poor Greg is used to degrees Fahrenheit, so when his relatives tell him it's going to be 30 degrees out, he has no idea whether to pack his swim trunks or his mittens. The formula for finding the Fahrenheit temperature from the Celsius temperature is:

$$F = \frac{9}{5}C + 32$$

... where F is degrees Fahrenheit and C is degrees Celsius. But that formula is a real pain to use mentally, so Greg's uncle David taught him to use a modified formula: F = 2C + 30. This might be a decent approximation, since 9/5 is roughly 2 and 30 is close to 32. How good is Uncle David's approximation? To figure that out, use the real formula and David's formula to find the Fahrenheit temperature for the following temperatures in degrees Celsius: 10, 20, 30, 40, and 50. For which temperature is David's formula most accurate? For which temperature is it least accurate?

What kind of variation do the two formulas represent?

Based on your answer to that last question, roughly sketch the two graphs on the same set of axes. (An accurate graph would need too wonky a scale to be useful, so don't bother with precision.) Based on the calculations that you did or the sketch that you made or both, for what Celsius temperature do you think David's formula will give the absolute best approximation? As the temperature continues to rise above 50 degrees Celsius, will David's formula get more or less accurate?

Do the following calculations:

7. 23.26 - 4.09

8. 541.2 - .06

9. (1.6)(.5)

10. (3.2)(2.7)

11. Here's a problem adapted from Norman D. Willis's book *False Logic Puzzles*:

Socrates was sentenced to death for corrupting the minds of the youth of Athens in 399 BCE. According to Xenophon and Plato, who wrote much of what we know about Socrates's life, Socrates's followers were able to bribe the prison guards to let him escape. Socrates elected not to leave, as he believed both that accepting his fellow citizens' judgment of his guilt was his duty as an Athenian and that escaping would indicate a fear of death, which no true philosopher has. Let us imagine, though, the scene when the authorities discovered the cell door left open by the bribed guards. The four men assigned to Socrates's prison make the following statements:

Guard A: I did not leave the door open. C did it.
Guard B: I was not on duty that night. A was.
Guard C: B was on duty that night. But I did hope Socrates would escape.
Guard D: I didn't leave the door open. I am not surprised that Socrates didn't try to escape.

You can assume there is no conspiracy among the guards; the man who was on duty is the one who left the door open. If three of the eight statements are true and five are false, which guard is guilty?

4
THE CONIC SECTIONS

1 FRACTIONAL EQUATIONS

In this chapter, you'll be working with fractional equations and looking at how they relate to geometry. I'd like you to start off by trying your hand at a truly classic problem — it might be as many as three thousand years old. It's okay if you don't solve it. Spend at least ten minutes working on it, either alone or with a group, using whatever techniques you can come up with. Later in this lesson, I'll walk you through an algebraic solution to it.

1. **Diophantus was a mathematician in ancient Greece. We don't know exactly when he lived, but we do know how long he lived because one of his admirers made up a riddle about his life. It goes like this: one-sixth of Diophantus's life was spent in childhood. One-twelfth of his life was spent as a youth. After that, a seventh of his life passed before he was married. Five years later, he had a son. Diophantus's son lived to be half as old as his father, and Diophantus died four years after his son. How many years did Diophantus live?**

Before we go any further, I'm going to ask you to do a little review work on algebraic fractions. You worked with them quite a bit in *Jousting Armadillos*, but it's been a little while.

First of all, it is absolutely crucial to remember that fractions and division are very closely related. Indeed, I'd say that they're really the same thing. For instance, it is possible to think about the mathematical expression $\frac{3}{5}$ in two ways. You can think of taking a whole thing

(an apple, a pie, the mass of the earth, or whatever you like), splitting that thing into five pieces, and keeping three of those pieces. This is what we usually mean when we read that expression as "three-fifths." You can also imagine taking three whole things and splitting that whole set into five equal groups. This is usually what we mean when we read that expression as "three divided by five." The point is that *either way of interpreting the expression $\frac{3}{5}$ is perfectly valid and, ultimately, they mean the same thing.*

Now for a quick review of how to work with fractions. These problems will probably be pretty easy for you, but it won't kill you to do them. Remember that in order to add or subtract fractions you need to find a common denominator. Multiplying fractions is very simple: just multiply the numerators and the denominators (and, if necessary, simplify the result). To divide fractions, you first find the reciprocal of the fraction that you're dividing by and then multiply it with the fraction to be divided.

Simplify the following expressions:

2. $\frac{3}{5} + \frac{7}{5}$

3. $\frac{8}{11} + \frac{2}{8}$

4. $\frac{4}{10} - \frac{4}{7}$

5. $\frac{6}{12} + \frac{3}{4}$

6. $\frac{12}{42} - \frac{7}{21}$

7. $\frac{3}{8} \cdot \frac{4}{6}$

8. $\dfrac{13}{19} \cdot \dfrac{38}{39}$

9. $\dfrac{6}{17} \div \dfrac{9}{34}$

10. $\dfrac{5}{14} \div \dfrac{15}{28}$

11. $\dfrac{7}{16} \div \dfrac{3}{64}$

Excellent. Now, as you learned in the last textbook, working with fractions that include variables is *exactly the same as working with fractions that don't include variables*. If you are comfortable working with variable-less fractions, working with variable-ful fractions is quite simple. ("Variable-less" and "variable-ful" are definitely not official math vocabulary, but I hope you see what I mean.)

If you can see that $\dfrac{2}{5} \cdot \dfrac{3}{4} = \dfrac{6}{20} = \dfrac{3}{10}$ and you are willing to believe that $(2x)(3) = 6x$, then I hope it's pretty clear that $\dfrac{2x}{5} \cdot \dfrac{3}{4} = \dfrac{6x}{20} = \dfrac{3x}{10}$. And since $(2x)(3x) = 6x^2$, then $\dfrac{2x}{5} \cdot \dfrac{3x}{4} = \dfrac{6x^2}{20} = \dfrac{3x^2}{10}$. In other words, multiplying variable-ful fractions is exactly the same as multiplying variable-less fractions. As I say, you learned all this in *Jousting Armadillos*, but it's been a while, so practice on these problems.

Simplify the following expressions, and don't be fooled into thinking that you can simplify them further than is possible. For example, $\dfrac{2x+y}{5}$ and $\dfrac{x^2+3x}{2}$ are both simplified as much as they can be.

12. $\dfrac{22x^2}{77} + \dfrac{4x}{11}$

13. $\dfrac{15y}{18} - \dfrac{3y}{9}$

14. $\dfrac{13}{m^4} + \dfrac{9}{m}$

15. $\dfrac{8}{x+2} - \dfrac{4}{3x+6}$

16. $\dfrac{5(x+3)}{6} - \dfrac{8}{18}$

17. $\dfrac{x}{4} - \dfrac{4}{x}$

18. $6x\left(\dfrac{11}{3x}\right)$

19. $\left(\dfrac{5}{x-3}\right)\left(\dfrac{3x}{2}\right)$

20. $\dfrac{4y^5}{3x^2} \div \dfrac{5y^8}{8y^2}$

Before you go on, make sure that you've checked the solutions to those problems with your fellow students and/or teachers. When you're confident that you've got it down, you can proceed.

All the work that you've done with algebraic fractions so far has involved simplifying expressions. Recall that the key difference between an expression and an equation is that an equation has an equal sign in it. $\frac{2x}{5} + \frac{3x}{4}$ is an expression; $\frac{2x}{5} + \frac{3x}{4} = 46$ is an equation. When you're working with expressions, all you can do is change their form — hopefully to a simpler form. In the case of equations, you can find their *solutions* — the value (or values) of the variable that makes the equation true. You have all the skills that you need in order to solve fractional equations with one variable. (In fact, you did a little bit of this in the last textbook as well.)

Let's start with a pretty simple algebraic fractional equation: **two-thirds of x equals six.**

21. **Write the mathematical version of "two-thirds of x equals six."**

There are a few ways that you could have written the **two-thirds of x** part of that equation. For reasons that you'll soon see, writing it as $\frac{2x}{3}$ is far more common and probably a lot more useful than writing it as $\left(\frac{2}{3}\right)x$, $\frac{2}{3}x$ or $\frac{2}{3} \cdot x$.

22. **Explain why $\frac{2x}{3}$ and $\left(\frac{2}{3}\right)x$ are two ways of writing the same expression.**

(Hint: The best way to convince yourself of this is probably to try replacing x with several different values.)

Now let's look at three different approaches to solving the equation $\frac{2x}{3} = 6$.

23. **The first approach could be called the *intuitive approach* or the *unit-rate method*. If two-thirds of x is six, what's one-third of x? If one-third of x is the number that you just decided on, what is all of x?**

Check to make sure that the number you just came up with makes the equation

$\frac{2x}{3} = 6$ **true. Then try a few more intuitive-approach questions:**

a. $\left(\frac{1}{3}\right)x = 7$ **b.** $\left(\frac{2}{5}\right)x = 10$

c. $\left(\frac{3}{4}\right)x = 21$

This approach is very useful for quick mental calculations, but you may find that it's tricky to apply to more complicated equations.

24. The next approach will probably feel overly complicated for solving this problem, but it's very useful for solving more complicated equations. It's called the *cross-product* or *cross-multiplication method.* You used it in *Jousting Armadillos* to solve proportion problems, but we'll take it a bit further here. Before you use it, explain why the equation $\frac{2x}{3} = 6$ is equivalent to the equation $\frac{2x}{3} = \frac{6}{1}$.

The cross-product method relies on the fact that the cross products of the two fractions in an equation are equal. In other words, you can write a new, equivalent equation using the products of the numerators and denominators of the two fractions. To give you an example without variables, $\frac{1}{2} = \frac{2}{4}$ can be rewritten as $(1)(4) = (2)(2)$, or $4 = 4$.

25. Use the cross-product method to rewrite the equation $\frac{2x}{3} = \frac{6}{1}$ and solve that equation. If everything went right, you got the same solution as you did in the last problem, and since you've already checked that it makes the equation $\frac{2x}{3} = 6$ true, you don't need to check it again. But practice a few more of these:

a. $\frac{x}{3} = \frac{7}{1}$ 　　　　　　　　　　b. $\frac{x}{3} = \frac{8}{6}$

c. $\frac{x}{4} = \frac{20}{5}$

26. The third approach is the one that I think you'll find most useful for solving the Diophantus problem that you started the lesson with. It rests on the very simple principle that basically no one really likes to work with fractions. It's usually pretty easy to rewrite an equation so that you get rid of all the fractions in it.

In the case of the equation $\frac{2x}{3} = 6$, all you have to do is ask yourself what number you would need to multiply the fraction $\frac{2x}{3}$ by so that, when you simplify it, you'd get an expression that's no longer a fraction. Go ahead and ask yourself that question. And write down the answer. Now, simply multiply the equation $\frac{2x}{3} = 6$ by that number, *remembering that in order to get an equivalent equation, you need to multiply both sides of the original equation by the same thing.*

You should now have a simple equation without any fractions in it. When you solve that equation, you ought to come up with the same solution that you did for the last two problems. Make sure you've got the hang of it:

a. $15 = \dfrac{5x}{3}$

b. $\dfrac{7x}{2} = 28$

c. $8 = \dfrac{4x}{5}$

This third approach can be especially useful in solving an equation that contains multiple fractions with different denominators, like this one:

$$\dfrac{3x}{5} + \dfrac{2x}{4} = 11$$

Here's the question to ask yourself:

27. What single number can you multiply the entire equation $\dfrac{3x}{5} + \dfrac{2x}{4} = 11$ by in order to get rid of all the fractions?

28. Go ahead and multiply all three elements of the equation by the number you've chosen, solve the equation, and check to see that your solution makes the original equation true.

Try a few more with different denominators.

29. $\dfrac{x}{2} - \dfrac{2x}{8} = 4$

30. $2x - \dfrac{4x}{3} = \dfrac{20}{6}$

31. $\dfrac{21}{x} = \dfrac{24}{2x} - 1$

One more complication and then I'll give you some problems to practice on. Consider an equation like this one:

$$\dfrac{4x + 4}{2} = \dfrac{12x}{5}$$

You can solve this equation easily enough using, for example, the cross-multiplication method *as long as you remember to apply the Distributive Property when you multiply 5 by 4x + 4.*

32. Go ahead and solve $\dfrac{4x + 4}{2} = \dfrac{12x}{5}$ and check your answer.

Solve the following equations using whatever approach or combination of approaches you like. Check your solutions. Remember that the beauty of these single-variable equations is that you can always check your solutions on your own without consulting a teacher or a classmate or an answer book — just check to see whether the solution you've arrived at makes the original equation true. (In some cases the answers themselves will be fractions — but check those as well!)

33. $\dfrac{7}{3} = \dfrac{x}{9}$

34. $\dfrac{5x + 1}{3x - 2} = \dfrac{4}{2}$

35. $\dfrac{x + 5}{2} = \dfrac{4x - 2}{6}$

36. $\dfrac{10}{x - 3} = \dfrac{4}{x + 3}$

37. $\dfrac{4x + 1}{4x + 4} = \dfrac{4}{8}$

38. $\dfrac{x}{3} = \dfrac{x}{2} - \dfrac{7}{1}$

39. $\dfrac{5}{x} + \dfrac{20}{5x} = 3$

40. $\dfrac{x - 5}{5} + \dfrac{x + 7}{2} = \dfrac{2x}{10}$

41. $3x - \dfrac{1}{2} = 4x + \dfrac{11}{2}$

42. $5x + \dfrac{2}{3} = \dfrac{8x + 4}{2}$

Problems 43 and 44 are sets of simultaneous equations. Solve each set by combining the techniques from the last chapter with the techniques in this lesson. Remember that the solution to a set of simultaneous equations consists of an x-value and a y-value. Check to see that your values make the original equations true!

43. $\dfrac{x}{3} + \dfrac{y}{2} = 3$

   $\dfrac{x}{5} + \dfrac{y}{4} = 1$

44. $\dfrac{x + 2}{4} - \dfrac{y + 11}{2} = -3$

   $\dfrac{-x + 11}{3} - \dfrac{2y + 4}{2} = 4$

For the next two problems, remember that when you solve equations that have a squared variable, two solutions are possible. For example, if $x^2 = 25$, then x could be either **5** or **-5**. (By the way, one way of writing that **x** can be equal to either **5** or **-5** is $x = \pm 5$.)

45. $\dfrac{x + 4}{4} = \dfrac{x + 1}{x}$

46. $\dfrac{x - 18}{x} = \dfrac{x - 2}{-2}$

47. Create four equations like the ones that you've just been working with in Problems 33-42 for your classmates to solve. It should require more than one step to solve each of them. Make sure that they have whole-number solutions.

Now that you're well practiced with fractional equations, let's look back at the Diophantus problem. It can be solved using the methods you've been practicing. (Who knows — maybe you came up with this method yourself!) To minimize the amount of page shuffling you have to do, here's the riddle again:

Diophantus was an ancient Greek mathematician. We don't know exactly when he lived, but we do know how long he lived because one of his admirers made up a riddle about his life. It goes like this: one-sixth of Diophantus's life was spent in childhood. One-twelfth of his life was spent as a youth. After that, a seventh of his life passed before he was married. Five years later, he had a son. Diophantus's son lived to be half as old as his father, and Diophantus died four years after his son. How many years did Diophantus live?

In my opinion, the most challenging part of doing this kind of problem is writing the equation, not solving it. The first thing that you need to do is to decide what the variable you're going to use will stand for. As it happens, you could decide to have the variable (let's call it **x** — always the most popular choice) stand for any number of parts of the problem. For instance, **x** could stand for the amount of time that Diophantus spent as a child, or the amount of time that he was married. But the simplest thing to do is to let **x** stand for the thing we're ultimately looking for: the total length of Diophantus's life.

48. If x is the total length of Diophantus's life, write an expression for the amount of time that he spent in childhood. Write an expression for the amount of time he spent as a youth. Write an expression for the amount of time that passed between his being a youth and getting married. Write an expression for the amount of time that passed between his getting married and having a son. (This will be different from the last three expressions — in fact, it will be simpler.) Write an expression for the length of Diophantus's son's life. Write an expression for the amount of time that Diophantus lived after his son's death.

The last six expressions that you wrote each stand for a chunk of Diophantus's life, and none of the chunks overlap. (The time that he spent in childhood goes right up to the time that he spent as a youth, but he didn't spend any time as both a child and a youth, and so on.) Therefore, if you added up all of the chunks that those expressions stand for, what should the total be? Write the equation for this addition process.

You should now have an equation with a long string of expressions being added together to equal a single expression. You can solve this equation using the methods that you've been practicing. I suggest using the third approach that you learned, where you get rid of the fractions. Just make sure that you multiply the entire equation by whatever number you decide will change all of the fractional expressions to non-fractional expressions.

In order to check your solution, I'd say that the best thing to do is to go back to the wording of the puzzle itself and make sure that the solution you found for Diophantus's life is the same as the total of all of the pieces of his life. (This is really the same as checking that your solution works in the equation that you wrote — unless you wrote the equation incorrectly, which is why I suggest going back to the original problem.)

49. Write a *Note to Self* that shows *how to solve fractional equations with one variable*. Include a couple of examples, though they need not be as complicated as the Diophantus problem. Show how to use cross multiplication and how to multiply in order to eliminate fractions.

50. Here is a fractional puzzle that I find quite challenging: If a cat and a half can eat a fish and a half in a day and a half, how many fish would seven cats eat in a week and a half?

51. The following question comes from the Rhind Papyrus, written around 1650 BCE: If a certain number, two-thirds of it, half of it, and a seventh of it are added together, the result is 97. What is the number?

REVIEW

NO CALCULATORS! (You can do several of these mentally.)

1. 7.06 + 1.36
2. 427.12 + 1.09
3. 23.06 - 45.92
4. 541 - 0.06
5. 0.0154 - 0.003
6. (12)(0.4)
7. (12.9)(4.2)
8. (24.7)(0.02)
9. (0.03)(0.015)
10. 3 ÷ 1.5
11. 6.4 ÷ 1.6
12. 1.25 ÷ 0.05
13. 326.1 ÷ 0.03
14. 0.0183 ÷ 4

15. Go back and check your answers to Problems 1-14 with a calculator. For any you missed, find out what you did wrong by inspecting your work — look for regrouping errors or misplacement of the decimal. If that does not rectify the error, redo the problem. If you are still stymied by a problem, consult a classmate, teacher, parent, etc. Then make up two similar problems to practice. For those problems you just created, repeat this process of solving and checking until you are perfectly comfortable performing all four operations with decimals.

16. Simplify this crazy-looking expression: $(1 - \frac{1}{2})(1 - \frac{1}{3})(1 - \frac{1}{4})(1 - \frac{1}{5}) \ldots (1 - \frac{1}{99})$
(It's not as hard as it looks when you figure out the pattern!)

17. Sam Loyd was famous for the puzzles he made in the late 1800s and early 1900s. Here's a puzzle based on one of his:

What is the shortest path that the armadillo can take from Circle 1 to Circle 12 if he follows the lines and passes through every circle on the way? (As you might imagine, he can't ride a bike very well, so he can't make turns and needs to remount in every circle in order to change direction.) Copy the puzzle or ask your teacher to make some photocopies so you can try different solutions without marking up your book.

2 TWO-INTERCEPT FORM

One of the things that you're becoming quite adept at is manipulating equations — changing them from one form to another. In this lesson, I'll be asking you to take a look at a new form of two-variable equations.

First, a reminder of two forms that you're already familiar with. There's *slope-intercept form*, which we've also called **y-equals** form. In its generic version, it looks like this: **y = mx + b**. A specific example might be **y = 4x - 2**.

1. In what situations is slope-intercept form most useful?

Another form of two-variable equations that you've encountered quite often is *standard form*. The generic version of standard form is **ax + by = c**. A specific example might be **3x - 7y = 12**.

2. In what situations have you found standard form to be the most useful?

The form that you're going to take a look at now is called *two-intercept form*. Here's its generic version:

$$\frac{x}{a} + \frac{y}{b} = 1$$

3. Compare the generic version of two-intercept form with the generic versions of the other two forms. What similarities do you notice? What differences? (There are several differences, of course, but one of them is key.)

Here are some specific examples of equations in two-intercept form:

$$\frac{x}{5} + \frac{y}{3} = 1$$

$$\frac{x}{2} - \frac{y}{7} = 1$$

$$\frac{3x}{2} + \frac{2y}{7} = 1$$

Notice a couple of things about these equations. First of all, you can, of course, subtract the $\frac{x}{a}$ or the $\frac{y}{b}$ elements as well as adding them. This should come as no surprise: after all, subtracting is just adding a negative, and **y = -4x - 6** is a perfectly legitimate example of the form **y = mx + b**. However, for reasons that you'll soon discover, it's usually preferable to write an equation like $\frac{x}{2} - \frac{y}{7} = 1$ as $\frac{x}{2} + \frac{y}{-7} = 1$. Second, notice that the $\frac{x}{a}$ and $\frac{y}{b}$ elements can take forms such as $\frac{3x}{2}$ or $\frac{2y}{7}$.

4. Rewrite the expression $\frac{3x}{2}$ as a fraction times the variable x. As you learned in the last lesson, $\frac{3x}{2}$ is by far the more common way of writing the expression, but I want you to be aware that the expression you just wrote means exactly the same thing.

5. For practice, rewrite each of the following expressions as a fraction multiplied by a variable and as an English phrase. I'll do the first one for you.

Rational Expression	(Fraction)(Variable)	English Phrase
$\frac{5x}{8}$	$\left(\frac{5}{8}\right)(x)$	five-eighths of x
$\frac{2x}{9}$		
$\frac{9x}{2}$		
$\frac{-4x}{15}$		
$\frac{-8x}{7}$		
$\frac{x}{4}$		

Moving between the various forms of two-variable equations is not particularly tricky — you have all of the tools that you need in order to do it. The one thing to be aware of is that in two-intercept form one side of the equation is always equal to one. In order to achieve that situation, you will need to do some division (or, in some cases, multiplication by the reciprocal).

6. On the following page you will be given equations in slope-intercept form, standard form, or two-intercept form. Convert each equation from its current form into the other two. Pay close attention to the methods you employ to manipulate these equations, because you will be asked to explain your methodology fully after this problem set. All three forms have been provided for the first two equations — practice on those to ensure that you are manipulating the equations correctly.

Slope-Intercept	Standard	Two-Intercept
$y = -x + 2$	$x + y = 2$	$\frac{x}{2} + \frac{y}{2} = 1$
$y = \frac{-4x}{3} + 4$	$4x + 3y = 12$	$\frac{x}{3} + \frac{y}{4} = 1$
		$\frac{x}{2} + \frac{y}{3} = 1$
		$\frac{x}{4} + \frac{y}{4} = 1$
$y = -2x - 4$		
$y = 3x + 9$		
	$5x - 3y = 15$	
		$\frac{x}{7} + \frac{y}{-21} = 1$
	$4x + 5y = 20$	
$y = 6x + 18$		
		$\frac{x}{9} + \frac{y}{3} = 1$
$y = -5x - 20$		
	$10x + 5y = 30$	
	$18x - 9y = 72$	
		$\frac{x}{5} + \frac{y}{-8} = 1$
$y = -4x - 3$		
		$\frac{x}{4} + \frac{y}{5} = 1$

7. Go back over the problems that you just did. Can you devise a step-by-step "recipe" for moving from the generic version of one form to another? In other words, is there a method that will always work to go from, say, $y = mx + b$ to $\frac{x}{a} + \frac{y}{b} = 1$? See if you can generalize a method for going from each form to each of the other forms. That should be six methods overall.

Great. You have quite a bit of practice changing from one form of equation to another, and you can probably transform just about any two-variable equation into two-intercept form. Now ask me why you should bother. Go ahead, I dare you.

You should *always* ask why you ought to bother learning something. Sometimes the answer will be something like, "Trust me: you'll need this later on," and if you do trust the person who's talking, then go ahead and accept that reason. I hope, though, that you'll usually get a more satisfactory answer. In the case of two-intercept form, the answer is that it's quite useful for doing one specific thing.

Let's consider the equation $\frac{x}{4} + \frac{y}{5} = 1$.

It's a two-variable equation in which neither of the variables is squared or anything fancy like that, so I hope you're willing to believe that its infinite solutions can be represented graphically by a straight line. (Also, you just saw that it's possible to change two-intercept form into **y-equals** form and you *know* those are straight lines.)

8. Suppose you wanted to find the y-intercept of that line. When the line crosses the y-axis, what does the value of x have to be? Or, in other words, when *any* line crosses the y-axis, what will be the x part of the coordinate (x, y)?

9. Rewrite and simplify the equation $\frac{x}{4} + \frac{y}{5} = 1$ using the value of x that you decided on.

10. Look at your new equation. In order to make it true, what number needs to replace the y to make the left side of the equation also equal one? That number is the y-intercept of the equation.

11. We haven't talked much about the idea of an x-intercept, but nearly every straight line has one — as you might guess, it's the point where the line crosses the x-axis. In order to find the x-intercept of $\frac{x}{4} + \frac{y}{5} = 1$, you simply have to do the steps you just did, only this time replace y in the equation above with the value that it has when the line crosses the x-axis. Go ahead and take those steps to find the x-intercept.

If you compare the values that you just found for the **x-** and **y-intercepts** of $\frac{x}{4} + \frac{y}{5} = 1$ with the equation itself, I hope you can see what the virtue of the two-intercept form is: it allows you to find the **x-** and **y-intercepts** of an equation very, very quickly. And, of course, if you know two points that lie on a straight line, it's very easy to graph that line. (One of the fundamental truths of geometry is that "two points determine a line.")

It should look like the graph below, right?

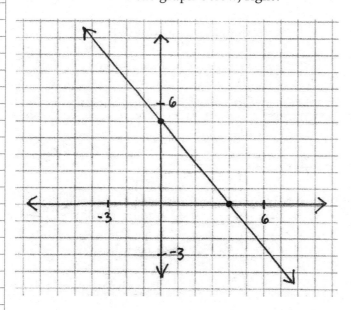

Among mathematicians, two-intercept form is often revered as the easiest form of a linear equation to graph. For this reason, I'd like you to try to mentally "see" the geometry of the line prior to putting it on a graph. Before you graph the following equations, pause and think about each one for just ten seconds while you visualize both the **x-** and **y-intercepts**.

Picture each graph in your head, then draw it.

12. $\frac{x}{3} + \frac{y}{5} = 1$ 13. $\frac{x}{-2} + \frac{y}{4} = 1$

14. $\frac{x}{6} + \frac{y}{-3} = 1$ 15. $\frac{x}{1} + \frac{y}{7} = 1$

For the next two problems, don't graph them — just picture them in your head and tell whether their slopes will be positive or negative.

16. $\frac{x}{-5} + \frac{y}{2} = 1$ 17. $\frac{x}{1} + \frac{y}{1} = 1$

All the previous problems have had their fractions separated by an addition sign, but many of the ones following do not. There's a simple way to change this. First, you will have to accept that these three expressions are all equal:

$$-\frac{y}{6} \qquad \frac{-y}{6} \qquad \frac{y}{-6}$$

18. To convince yourself that $-\frac{y}{6} = \frac{-y}{6} = \frac{y}{-6}$, try replacing y with any number.

Do you get the same value for all three expressions?

Rewrite the following equations so that they don't involve subtraction and so that their x- and y-intercepts are clear. Then picture them and say whether their slopes are positive or negative. (One of these problems will be *very* difficult... when you get to that one, explain why it's so difficult.)

19. $\frac{x}{3} - \frac{y}{6} = 1$ 20. $\frac{x}{-2} - \frac{y}{-5} = 1$

21. $\frac{-x}{6} - \frac{-y}{2} = 1$ 22. $\frac{-x}{-7} - \frac{-y}{-2} = 1$

23. $\frac{x}{-4} + \frac{y}{0} = 1$

24. Now for a slightly more difficult one. Consider the equation $\frac{2x}{3} + \frac{2y}{5} = 1$. The x- and y-intercepts of this one may not be so obvious at first glance, but actually they're easy to find as well — they just aren't whole numbers. In order to see what they are, I think it's best to re-imagine $\frac{2x}{3}$ as $\frac{2}{3}(x)$. In that case, the equation at the x-intercept (when y is zero) would be $\frac{2}{3}(x) = 1$. So the x-intercept must be the thing that you would have to multiply $\frac{2}{3}$ by in order to get one. What is that?

Use the same method to find the y-intercept of the equation $\frac{2x}{3} + \frac{2y}{5} = 1$. Graph it!

(Hint: You'll get a more accurate graph if you use a scale of two graph paper squares to one unit instead of the usual one to one.)

I hope your graph looks something like this:

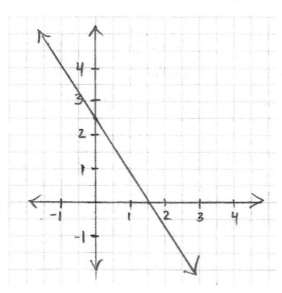

Find the x- and y-intercepts of the following — you don't have to graph them, but please express each answer as a common fraction. (Simplify them, but you don't need to change improper fractions into mixed numbers.)

25. $\dfrac{2x}{3} + \dfrac{3y}{1} = 1$

26. $\dfrac{3x}{4} + \dfrac{2y}{5} = 1$

27. $\dfrac{12x}{4} + \dfrac{4y}{8} = 1$

28. $\dfrac{-2x}{5} + \dfrac{9y}{-4} = 1$

29. $\dfrac{8x}{-3} - \dfrac{-3y}{3} = 1$

30. $\dfrac{3x}{3} + \dfrac{4y}{4} = 1$

31. Curious as it may seem, you can actually think of an equation like $5x + 3y = 1$ as being in the form $\dfrac{x}{a} + \dfrac{y}{b} = 1$. Using the same reasoning as you did for the last few problems, find the x- and y-intercepts of the equation $5x + 3y = 1$.

32. It's time to write a **Note to Self** about **two-intercept form**. Be sure that the Note includes the generic version of two-intercept form and that it explains how to use two-intercept form to find the x- and y-intercepts of an equation.

33. Earlier in this lesson you practiced moving between various forms of equations, but in fact, there are certain equations that are impossible to write in two-intercept form. Try to write an equation in y-intercept form (or standard form) that cannot be written in two-intercept form and explain why it can't. (Hint: Look back at Problem 23.)

34. Linus and Greg are workers on an assembly line producing thingamajigs. The bosses are not pleased with the quality of their work: their frequent mistakes have led many customers to complain and several injuries have been sustained from poor attachments of the doohickeys to the whatchamacallits. In an attempt to eliminate mistakes, the bosses have instituted an incentive program. If zero mistakes are made during the day, Linus and Greg will each receive a bonus of $20. If they make five errors in production of the thingamajigs, they will receive no bonus at all. Assuming that this is a linear relationship, determine the x- and y-intercepts and coordinates, then express this incentive program as an equation in two-intercept form, letting x represent the number of mistakes made and y represent the bonuses earned.

35. Sarah's Glove Hut has recently analyzed its business plan and realized that there is a linear relationship between the outside temperature and the number of gloves it sells. When the temperature reaches 80°F, Sarah sells no gloves, but when the temperature outside dips down to 0°F, she sells 120 gloves per week. Create an equation in two-intercept form to express this relationship, telling what each variable stands for.

36. Next week the temperature is predicted to be 40°F. How many gloves does Sarah need to have in stock to be ready for her customers?

37. This fall she is opening a second location in Ulan-Bator, Mongolia, where the January average temperature is about -20°F. If the linear relationship between temperature and gloves sold holds true, how many gloves can she expect to sell in January?

Here's one last problem that's related to standard form, though it can be solved with one variable. I've seen a number of different variations on this problem; they always look deceptively simple and I always find them incredibly hard. (In fact, it was Greg who showed me how to do this — I couldn't have done it myself!) I'll give you the problem and let you wrestle with it and then I'll show you how to set it up step by step in Problems 39 - 42. If you've already figured out how to do it, you don't need to do those problems. Okay, so here it is:

38. There are two woodchucks named Ginger and Maybelline. Ginger can chuck a standard pile of wood in three hours and Maybelline can chuck a standard pile of wood in five hours. Working together, how long would it take them to chuck a single pile of wood?

Check to see that your answer to Problem 38 (if you got one) makes sense to you. For one thing, it had better take the two of them less time to chuck together than it would take either of them to chuck on her own. If your answer doesn't make sense or if you didn't get one...

39. Okay, let's consider one woodchuck on her own. If Ginger chucks for x hours, I think this equation is true:

$$\frac{x}{3} = \text{\# of piles she chucks}$$

Make sure you believe me: plug in some values for x and see that they give results you think are logical.

40. There's a similar equation that represents the number of piles that Maybelline can chuck in x hours. Write it and make sure that you believe it by plugging in x-values.

41. Now, based on the work that you've just done, I'd say that this is true when Ginger and Maybelline work together for x hours:

$$\frac{x}{3} + \frac{x}{5} = \text{total \# of piles they chuck}$$

See what happens when you plug 15 hours into that equation and whether it makes sense to you.

42. But the thing is, in the original problem you're trying to figure out how long they worked and you know how many piles they chucked together. So you should be able to plug in that number for "total # of piles they chuck" in Problem 41 and use the techniques from the last lesson to solve for x.

Okay, now you know how to do that kind of problem (and so do I!), so the next time you see one, you'll be able to nail it.

REVIEW

1. The sum of seven consecutive integers is 413. What is the first number? (Hint: Try calling one of the numbers x and writing an equation.)

2. The average of two numbers is 56. If one number is 10 more than the other, what are the two numbers?

3. Mr. Dale bought a table and 4 chairs. The table cost 3 times as much as each chair. If he spent $189 altogether, how much did he pay for each chair?

4. Imagine that a page of 12-point type contains 500 words, while a page of 10-point type contains 660 words. A student printed a report using 10-point type and the result was 25 full pages. If she prints the same report in 12-point type, how many pages will it contain?

This next set of problems is not technically review, but it has to do with an algebraic concept that I've often seen on standardized tests like the SSAT and the SAT. These problems don't fit into any of the regular lessons in this book, but I think you'll find it helpful to study them and so I'm inserting them here. The way these problems work is that they give you a definition for a symbol and then ask you to use that symbol. You know the definitions for tons and tons of symbols already. Here are four you're quite familiar with:
+ - • ÷

Those symbols tell you about the way that two or more numbers interact. **2 • 5** means "multiply **2** by **5**," and so on.

These particular symbol problems I'm talking about usually look something like this:

For all integers x, $*x = \left(\dfrac{x}{x-2}\right) + 1$

What they've done is to give you the definition of the symbol: ***5** would mean "divide **5** by **3**, then add **1**." Then they make you do calculations.

Try some, using the definition of *x above:

5. *4

6. *6

7. *1

8. **There is an integer for which * would give you an impossible calculation. What integer is that?**

You might also see these problems presented in another way. They might give you one or more examples of made-up symbols in use, like so:

2 △ 3 = 10
3 △ 5 = 16
4 △ 7 = 22
7 △ - 2 = 10

Then they would ask you to use their made-up symbol on two other numbers. In this case, you have to figure out what △ means.

Figure it out and use it to make the following calculations:

9. 2 △ 6

10. 5 △ - 3

I'll give you some more problems like this in future review sets.

11. **At a constitutional convention, some delegates wear three-cornered hats and some wear two-cornered hats. If there are three times as many three-corner wearers as two-corner wearers and 121 total hat corners, how many delegates of each type are there?**

12. **Here's a puzzle based on one by Alan Wareham:**

Divide the square into four pieces of equal size and shape. Each piece must contain a 1, a 2, and a 3.

3 FRACTIONAL EQUATIONS WITH EXPONENTS

In this lesson, I'm going to ask you to look at a particular type of fractional equation that you haven't encountered yet. At first glance, this type of equation will look similar to the two-intercept form of linear equations that you learned about in the last lesson. In fact, it is quite different.

Here's an example of the type of equation that I'm talking about:

$$\frac{x^2}{25} + \frac{y^2}{25} = 1$$

1. **What's the first thing that makes this equation different from the linear two-intercept equations you just dealt with? (Hint: Look at the title of this lesson.)**

The fact that the **x- and y-variables** have exponents (in this case, they're both squared) radically changes the nature of this equation. It's worth mentioning that although you've encountered equations with exponents before, such as $y = x^2 + 3$ (and you'll see more of that kind in the next lesson), this is the first time that you've dealt with an equation in which *both* of the variables have exponents. One consequence of this is that such equations are not functions. Remember, one of the things that define a function is that for each value of the independent variable (**x**) there can be only one value of the dependent variable (**y**). As you will shortly see, that is not the case for equations like the one we're dealing with here. You don't need to get too hung up on the fact that these kinds of equations don't qualify as functions, but you should tuck it in the back of your mind somewhere.

So let's investigate the equation I gave as an example above:

$$\frac{x^2}{25} + \frac{y^2}{25} = 1$$

Whenever you have a two-variable equation, you can graph it on a Cartesian coordinate plane as you've done before. In the case of this kind of equation, its graph is probably its most important and interesting aspect.

The best way to graph it is to start with its **x- and y-intercepts.**

2. **Set up a coordinate graph that goes up to 8 in the positive directions and to -8 in the negative directions.**

3. **Ask yourself, "What will the value of y be when x is zero?" (That's the y-intercept, right?) In order to answer this question, rewrite the equation with x being equal to zero. One of the parts of the equation will basically disappear and you'll be left with a one-variable equation. This equation takes two fairly easy steps to solve, but here's the key: *remember that equations like this one have two solutions*. (This has to do with the fact that a negative number times a negative number gives you a positive number.) Find the two solutions. These two solutions are the *two y-intercepts* of this equation. Mark them as points on your graph.**

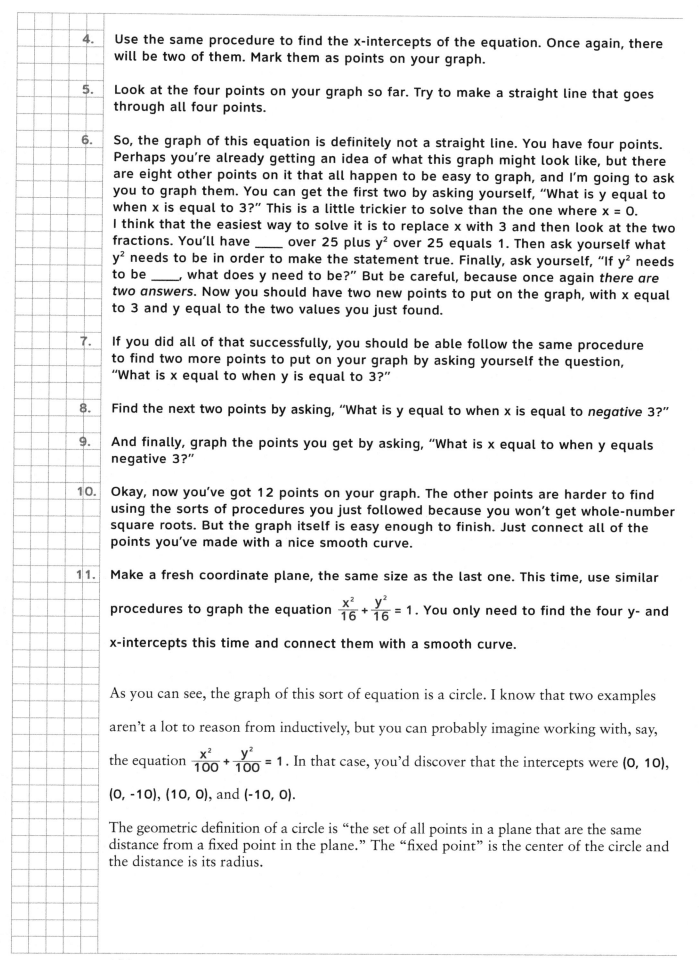

4. Use the same procedure to find the x-intercepts of the equation. Once again, there will be two of them. Mark them as points on your graph.

5. Look at the four points on your graph so far. Try to make a straight line that goes through all four points.

6. So, the graph of this equation is definitely not a straight line. You have four points. Perhaps you're already getting an idea of what this graph might look like, but there are eight other points on it that all happen to be easy to graph, and I'm going to ask you to graph them. You can get the first two by asking yourself, "What is y equal to when x is equal to 3?" This is a little trickier to solve than the one where x = 0. I think that the easiest way to solve it is to replace x with 3 and then look at the two fractions. You'll have ____ over 25 plus y^2 over 25 equals 1. Then ask yourself what y^2 needs to be in order to make the statement true. Finally, ask yourself, "If y^2 needs to be ____, what does y need to be?" But be careful, because once again *there are two answers*. Now you should have two new points to put on the graph, with x equal to 3 and y equal to the two values you just found.

7. If you did all of that successfully, you should be able follow the same procedure to find two more points to put on your graph by asking yourself the question, "What is x equal to when y is equal to 3?"

8. Find the next two points by asking, "What is y equal to when x is equal to *negative* 3?"

9. And finally, graph the points you get by asking, "What is x equal to when y equals negative 3?"

10. Okay, now you've got 12 points on your graph. The other points are harder to find using the sorts of procedures you just followed because you won't get whole-number square roots. But the graph itself is easy enough to finish. Just connect all of the points you've made with a nice smooth curve.

11. Make a fresh coordinate plane, the same size as the last one. This time, use similar procedures to graph the equation $\frac{x^2}{16} + \frac{y^2}{16} = 1$. You only need to find the four y- and x-intercepts this time and connect them with a smooth curve.

As you can see, the graph of this sort of equation is a circle. I know that two examples aren't a lot to reason from inductively, but you can probably imagine working with, say, the equation $\frac{x^2}{100} + \frac{y^2}{100} = 1$. In that case, you'd discover that the intercepts were (0, 10), (0, -10), (10, 0), and (-10, 0).

The geometric definition of a circle is "the set of all points in a plane that are the same distance from a fixed point in the plane." The "fixed point" is the center of the circle and the distance is its radius.

12. Looking at the equations we've been working with, how can you tell from an equation, without graphing it, what the radius of the circle it represents is? How can you tell from the graphs that these sorts of equations are not functions?

Graph the following equations. If necessary, manipulate them into the correct form.

13. $\dfrac{x^2}{9} + \dfrac{y^2}{9} = 1$

14. $x^2 + y^2 = 49$

15. $x^2 + y^2 = 6.25$

16. $\dfrac{x^2}{12.25} + \dfrac{y^2}{12.25} = 1$

17. What is the radius of a circle with the equation $x^2 + y^2 = 1$?

18. What is the equation of a circle with a radius of 12?

All right, now I'm going to ask you to examine a variation on this new kind of equation:

$$\dfrac{x^2}{16} + \dfrac{y^2}{36} = 1$$

Notice that, because the denominators of the two fractions in this equation are different, it's not quite clear what the radius of the circle it represents would be. In fact, it doesn't represent a circle, although it does represent something similar.

19. This new equation can be graphed by the same method you used for the equations of circles. Your coordinate plane can be the same size as the ones you used previously (extending to 8 or -8 in each direction). This equation has two y-intercepts, which you can find by making x equal zero. Find those intercepts and put them on your graph.

It also has two x-intercepts. Use the same technique to find those and graph them. There are a few other points that you could find, but I think you can imagine what this shape is like already. Go ahead and connect your four points using smooth curves. The shape should not have any unsightly bulges anywhere.

The shape you've just made looks like a slightly squashed circle. You might call it an oval, but its mathematical name is *ellipse*. Notice that the top and bottom halves of your ellipse are mirror images of each other, as are the right and left halves.

Graph the following ellipses. If necessary, convert them to the correct form.

20. $\dfrac{x^2}{25} + \dfrac{y^2}{4} = 1$

21. $\dfrac{x^2}{4} + \dfrac{y^2}{1} = 1$

22. $\dfrac{x^2}{9} + \dfrac{y^2}{4} = 1$

23. $16x^2 + 9y^2 = 144$

An ellipse does not have a single radius as a circle does. Instead, an ellipse is characterized by two distances: the *major axis*, between the two most distant edge points, and the *minor axis*, between the two closest edge points. There's a drawing of this on the next page.

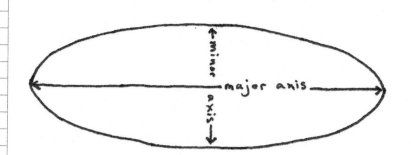

24. Look back at the ellipses that you graphed. How is it possible to tell, just from looking at the equation, what the lengths of the major and minor axes of the ellipse will be?

25. What is an equation for an ellipse with a major axis of five and a minor axis of three?

26. There are two possible answers to Problem 25. What's the other one? How would the graphs of the two ellipses that go with those two equations be different?

27. What are the lengths of the two axes of the elliptical equation $\frac{x^2}{12.25} + \frac{y^2}{6.25} = 1$?

The generic form for an equation of circle or an ellipse is $\frac{x^2}{a^2} + \frac{y^2}{b^2} = 1$.

If you want to, you can actually think of a circle as a special kind of ellipse where **a = b**.

Earlier in this lesson, I told you the technical, geometrical definition of a circle. (Just to remind you, I said that a circle was *the set of all points in a plane that are the same distance from a fixed point in the plane.*) The technical definition of an ellipse is a little different, and I'm not going to tell it to you — instead, I'll let you figure it out.

28. To do this problem, you'll need some supplies: two pushpins, some string, a pencil or pen, some paper, and a piece of cardboard — the thick kind that they make boxes out of is best because the pushpins need to stick into it. I'd also suggest doing this problem with a set of partners, just because it will be easier.

Get a nice, flat piece of cardboard. I'd say that it ought to be about twelve inches square, but you don't need to obsess about the dimensions.

Put a sheet of blank paper on top of the cardboard and stick the two pushpins through the paper and into the cardboard, not too near the edges and maybe three or four inches apart. (Again, don't worry too much about the distance.)

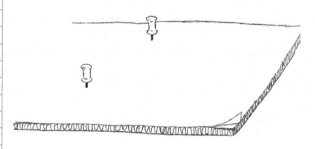

Tie the string into a loop. The loop needs to fit around the two pushpins with some slack in it (though not a huge amount).

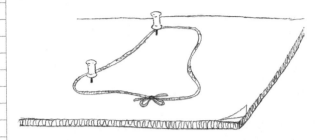

Put the loop around the pushpins and put the pencil or pen inside the loop. Stretch the loop tight so that the two pushpins and the point of the pencil make a triangle. Now draw on the cardboard, moving the pencil so that the string is always tight and forms an ever-changing triangle. Move the pencil until it comes back to the point where it started. If the pencil goes off the cardboard, your loop is too long. Re-tie it so that it is shorter and try again.

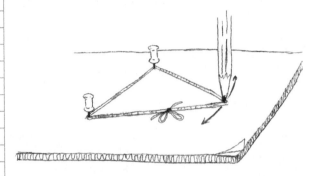

If everything went according to plan, you should have ended up drawing an ellipse on the paper.

Draw a few different ellipses. There are two ways to change your ellipses: you can change the length of the loop or the distance between the pushpins. See what happens when you do those two things.

If you think carefully, you can use the ellipses you've drawn to come up with the geometric definition of an ellipse. First you need to know that the points marked by the pushpins are called the *foci* of the ellipse. (Foci is pronounced "foh-sigh" and is the plural form of *focus*.)

Now, a circle is the set of all points that are the same distance from the center of the circle. Clearly, the points on an ellipse are not equal distances from either of the foci. But they do have a relationship to the foci. Make one more ellipse, going very slowly. Think about the triangle that is formed at every single point as you go around the ellipse. Notice that the edge of the triangle that runs between the two pushpins is always the same length and that the lengths of the other two edges change. But even though their lengths change, something about those two lengths or distances is always the same. What is it? (Hint: The total length of the loop never changes.)

Once you have figured out what is always true about those two lengths, use it to write a geometric definition of an ellipse. Be sure to check it with your teacher (as well as your classmates) — he or she may want to help you choose your words to make your definition clear.

From the days of the ancient Greeks — who, as you may know, had figured out that the earth orbits around the sun — it was believed that the earth (and the other planets) traveled in circles. During the time known as the Renaissance, artists and scientists (who were often the very same people) began to rediscover and explore some of those ancient ideas, many of which had been preserved and furthered by Muslim scholars. Those scientists and artists began to observe the world very, very closely, and some of them turned their attention to the sky. One particular man named Tycho Brahe, who lived from 1546 to 1601, recorded the movements of the sun, the moon, the stars, and the planets. His were by far the most thorough such recordings that had ever been made.

The problem with the observations that Brahe made is that they didn't fit with the idea of the planets moving in circular orbits. But most people were so set on this 2,000-year-old idea that they simply could not imagine giving it up. In order to make the circular idea match the actual observations, they moved from the theory of simple circles, like this…

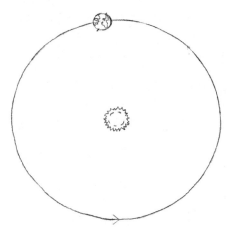

… to imagining that the planets moved in circles on circles, like this…

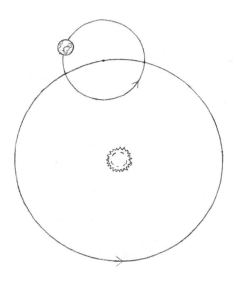

... or even circles on circles on circles, like this...

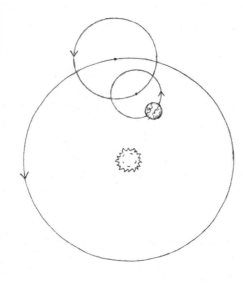

... and so on. One student of Brahe's named Johannes Kepler couldn't bring himself to believe in the circles on circles and he spent twenty years of his life studying the observations that Brahe had made, making new guesses as to how the planets moved, checking those guesses, finding that they failed, guessing again, and on and on. The thing that truly separated Kepler from most people who had gone before him — and, some people say, the way that Kepler ultimately changed how scientists around the world work — was that he refused to believe a theory that did not line up with actual, observed facts. Ultimately, through a combination of luck, inspiration, and perseverance, Kepler hit upon the explanation that *did* fit with the facts, the one that is still accepted today: the planets do *not* travel around in circles with the sun at the very center. They travel in ellipses with the sun at one focus of the ellipse and empty space at the other focus, like this:

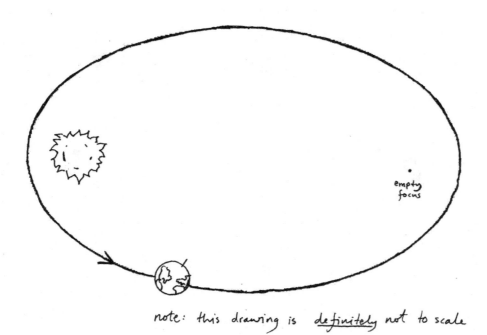

note: this drawing is <u>definitely</u> not to scale

I'm not going to ask you to do any problems based on actual planetary orbits because the calculations involved are pretty mind boggling. We'll stick with equations that give you more obvious square roots to work with for the time being. But I would like you to realize that the math you're learning right now is still helping astronomers and physicists explore the universe today.

As you may have noticed, the circles and ellipses that you graphed in this lesson were all centered on the origin. It is possible to create a circle or an ellipse that isn't centered on the origin, but you need to use a more complex form of the equation $\frac{x^2}{a^2} + \frac{y^2}{b^2} = 1$.

In the case of circles, the more complex form looks like this:

$(x - h)^2 + (y - k)^2 = r^2$

... and is called *the equation of a circle in standard form.*

When you have an equation in that form, **h** and **k** give you the **x-** and **y-coordinates** of the center of the circle and **r** gives you its radius. So the graph of $(x - 2)^2 + (y - 3)^2 = 9$, for instance, looks like this:

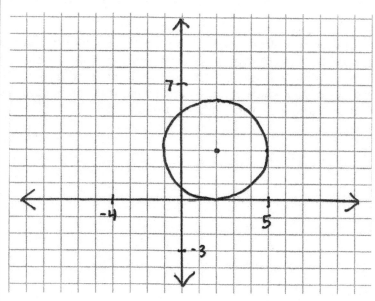

Graph the following equations:

29. $(x - 4)^2 + (y - 2)^2 = 4$

30. $(x + 2)^2 + (y - 5)^2 = 1$ (In this case, you need to remember that adding 2 is the same as subtracting negative 2.)

31. It may be hard to see that $\frac{x^2}{a^2} + \frac{y^2}{b^2} = 1$ and $(x - h)^2 + (y - k)^2 = r^2$ are two versions of the same thing, but it's actually not that hard to prove. To do so, first write down the equation $(x - h)^2 + (y - k)^2 = r^2$. If you used that formula for a circle with its center at the origin, what would the values of h and k be?

Plug those values into the equation $(x - h)^2 + (y - k)^2 = r^2$ and simplify it. What do you need to divide both sides of this simplified equation by in order to make it look more like $\frac{x^2}{a^2} + \frac{y^2}{b^2} = 1$? Go ahead and divide both sides by that thing.

Since, in the case of a circle, both a and b in the equation $\frac{x^2}{a^2} + \frac{y^2}{b^2} = 1$ stand for the circle's radius, I hope you can now see that $\frac{x^2}{a^2} + \frac{y^2}{b^2} = 1$ and $(x - h)^2 + (y - k)^2 = r^2$ really are two versions of the same thing.

32. It's time for either one or two **Notes to Self**. It's up to you whether you want to put *circles* and *ellipses* in the same Note or different Notes. In either case, you should include the geometric definitions of both shapes and how to graph each shape based on its equation. It's up to you whether you want to use the more complex circle equations that include h and k or the simpler version.

REVIEW

1. $6\frac{1}{4} - \left(-\frac{2}{3}\right)$

2. $\frac{2}{3}\left(\frac{1}{4} - \frac{1}{12}\right) \div \frac{1}{2}$

3. Express $8\frac{5}{8}$ as a decimal rounded to the nearest hundredth.

4. Express $\frac{6}{15}$ as a percent.

5. Half of Mona's money equals 3/5 of Nora's money. If Mona has $15 more than Nora, how much money do they have altogether? (Hint: You can do this as a pair of simultaneous equations.)

You can use a calculator on the next two problems, but write out the steps so that your teacher can read them easily. For example, you might write:

$$\frac{13 \text{ donuts}}{1 \text{ box}} \cdot \frac{144 \text{ boxes}}{1 \text{ crate}} = \frac{1{,}872 \text{ donuts}}{1 \text{ crate}}$$

6. Donuts are shipped from the bakery in large delivery trucks. The donuts are packed in boxes of thirteen donuts each (a baker's dozen); boxes are placed in crates that hold 144 boxes (a gross) each. The delivery trucks are packed 8 crates wide, 22 crates deep, and 17 crates high. How many donuts are in 10 truckloads?

7. Linus raised 80 weasels; he then entered into a series of business transactions. He traded all the weasels for aardvarks at an exchange rate of 5 weasels for 8 aardvarks. Next, he exchanged all the aardvarks for alligators at a rate of 4 aardvarks for 2 alligators. The alligators weighed 250 pounds each, and he sold all of them at a market price of $55.00 per 100 pounds. How much money did he make? How much did he make per weasel? Does weasel-raising seem like a lucrative enterprise in this economy or what?

8. 20% times 30% equals what percent?

9. What is 50% of 1/50?

10. The track team at Arbor School plans to sell T-shirts to raise money for new equipment. In order for the team to make a profit, the income from selling the T-shirts must be greater than the cost of making the T-shirts. At Rollman Silk-Screen, printing costs are $0.80 per shirt and the cost for each T-shirt is $3.75. The shop also charges a $125 fee per order for creating the unbelievably excellent silk-screen design.

   a. Write an equation for the total cost, y, of purchasing and printing x shirts.

   b. Copy and fill in the chart below. (You can use a calculator and round costs to the nearest whole cent for the rest of this problem.)

Number of Shirts	100	250	500	750
Total Cost				
Cost per T-shirt				

   c. If the track team orders 250 T-shirts and wishes to make a $500 profit, how much should they charge per T-shirt?

11. The made-up operation ~ is defined like this: $x \sim y = |x| - |y|$. Find the following:

   5 ~ 10
   3 ~ -3
   -3 ~ -3

12. Here are several examples of the made-up operation &:

3 & 7 = -11   2 & 1 = 0   3 & -2 = 7   -5 & 1 = -7

What is 10 & 2?

13. Graph the solutions to the following set of simultaneous inequalities:

$y - 3 \leq 2x$ $2x - 2y < 10$

14. Here's another puzzle from Sam Loyd.

Use a single cut, cutting only along the graph paper lines, so that the shape below is divided into two pieces that can be rearranged to form an eight-by-eight square:

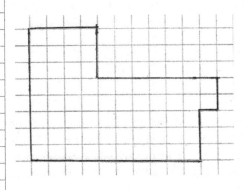

4 PARABOLAS

In the next two lessons, we're going to take a more detailed look at two curves that you've already encountered.

1. Find a ball (a tennis ball works well for this) or, if there's no ball available, any smallish object that you can toss from hand to hand. Stand up and toss the ball into the air from one hand to the other, watching the shape that it traces in the air. Do this several times. What shape does the path of the object seem to have?

You've probably noticed this before, maybe without wondering about it: any object thrown or otherwise propelled upward travels along a parabola. The fundamental reason for this is that the force of gravity pulls on an object so that the longer the object falls, the faster it moves. You don't need to worry too much about how exactly this works right now — you'll study it in detail when you take a physics class. It's just another example of the fact that mathematics can be used to represent fundamental truths about the universe.

I hope you'll remember from Chapter 2 that parabolas can be represented in algebra by equations that include exponents. The equation for the simplest parabola is $y = x^2$. You've seen this graph before, but now I'd like you to practice making it.

2. Make a coordinate graph that stretches from -5 to 5 on the x-axis and from -1 to 17 on the y-axis. Graph the equation $y = x^2$ by filling in this table and plotting the points:

x	-4	-3	-2	-1	0	1	2	3	4
y									

Now connect the points to form a smooth curve.

That is the most basic parabola. There are all sorts of ways to play with it, some of which we'll look at now.

3. Consider the equation $y = -x^2$. Before you graph it, make a hypothesis as to what it will look like. (It is a variety of parabola.) Then fill in a table as you did for Problem 2, with values for x from -4 to 4. Finally, make a coordinate graph that is the right size to fit those points, plot them, and connect them with a smooth curve.

4. This time consider the equation $y = 2x^2$. Based on what you've found already, predict what its shape will be. (Again, it's a variety of parabola.) This time, I'd recommend making a table with values for x from -3 to 3. Once you've made the table, graph and connect the points. Compare this graph to your graph of $y = x^2$ and explain how they're different.

5. Based on the work you've done so far, it should be pretty easy to predict the shape of $y = -2x^2$. Make your prediction, your table, and your graph.

6. Now consider the equation $y = \frac{x^2}{2}$. As you've been doing, predict, table, and graph. This time your table can go from x = -5 to x = 5. Some of your y-values will not be whole numbers, but I know you can handle that.

7. You should also be able to tackle $y = \frac{-x^2}{2}$ now. Go for it.

8. Remember that, in the equation $y = 2x^2$, the "2" is called the coefficient of x^2.

    What is the coefficient of x^2 in the equation $y = -2x^2$? In the equation $y = \frac{x^2}{2}$?

    (This one's a little trickier.) How about in the equation $y = x^2$? (Yes, there's a coefficient! It just doesn't normally get written.) And in the equation $y = -x^2$?

    What's the coefficient of x^2 in $y = \frac{-x^2}{2}$?

9. I hope you have a pretty good idea of how the coefficients in these equations affect their graphs. For this problem, I want you to draw a pair of axes for a coordinate plane, but don't worry about making accurate graphs — just make sketches and *do them all on one pair of axes so you can compare them*. On your coordinate plane, sketch graphs for the following equations and label them with their letters:

    a. $y = 3x^2$ b. $y = \frac{x^2}{3}$

    c. $y = 5x^2$ d. $y = \frac{x^2}{5}$

    e. $y = -4x^2$ f. $y = \frac{-x^2}{4}$

10. Write a set of rules explaining how the coefficient affects the graph of a parabola. You can use terms like "more steep" and "less steep" for your parabolas even though talking about the "steepness" of a parabola is actually a little bit complicated. (That's because the "steepness" or "slope" of a parabola, unlike the slope of a straight line, is not a constant. But you won't really need to worry about that until you study calculus.)

11. Now I'll ask you to look at a different way to alter the graph of a parabola. Just as you did for Problems 2 through 7, first make a prediction of what you think the graph will look like, then a table, then the graph itself. The table can go from x = -4 to x = 4. This time the equation is $y = x^2 + 1$.

12. Great. Now try the same steps for $y = x^2 - 2$.

13. Just so you're convinced, try $y = x^2 + 3$ and $y = x^2 - 1$. You can go straight to graphing the points if you're comfortable doing that. You can graph them on the same set of axes if you like.

14. In the equation $y = x^2 + 3$, the "3" can be called the *constant*. Write a rule for the way that the constant affects the graph of a parabola. (It will help you to write this rule if you know that the very tip of a parabola is called its *vertex*.)

Graph the following parabolas by combining the two rules you've written so far:

15. $y = -x^2 + 2$

16. $y = -2x^2 - 4$

17. $y = \dfrac{x^2}{2} + 3$

18. All right. One last way to play with the graph of a parabola. This time, try making a prediction, a table, and a graph for $y = (x + 1)^2$. (Yes, it's still a parabola. Frankly, I'll be surprised if your prediction is precisely right unless you've encountered these before.) Your table should go from -5 to 3.

19. Now try the same steps for $y = (x - 2)^2$. This time your table should go from -2 to 6.

20. I know that's only two examples, which is not a great foundation for a piece of inductive reasoning, but go ahead and make a rule for how equations like the ones in Problems 18 and 19 affect the graphs of parabolas. Unfortunately, as far as I know, there is no generally accepted name for the "1" in an equation like $y = (x + 1)^2$. That means that you can go ahead and name it whatever you want. If you become a famous mathematician one day, perhaps your name for it will be used in future algebra textbooks.

21. Based on the rule that you just wrote, make graphs for the following equations. (At this point, you can skip the predictions and the tables and go straight to the graphs if you like.)

    $y = (x - 3)^2$
    $y = (x + 6)^2$

22. You may be wondering whether the rules that you've made can be combined in making parabolas. They can. In order to show this, try graphing the equation $y = (x - 1)^2 + 2$. Make a prediction first. (I'll bet that your prediction is correct.) As you can probably predict, your table should go from -3 to 5.

23. Try $y = (x + 4)^2 - 3$ and $y = (x - 2)^2 - 2$.

24. Finally, in order to show that all three rules can be combined, try $y = -2(x - 1)^2 + 1$. Make your prediction first. This time your table can go from -2 to 4.

25. Write an equation for a parabola of ordinary "steepness" that opens downward with its vertex at (2, -3). Graph your equation to see whether you were correct.

For the following equations, give the parabola's vertex and tell whether it opens upward or downward. You do not need to graph them.

26. $y = (x - 3)^2 + 4$

27. $y = 2(x + 3)^2 + 5$

28. $y = -3(x + 9)^2 - 2$

29. $y = -\dfrac{1}{4}(x - 2)^2 - 3$

30. $y = 5(x + 1)^2 - 6$

In an earlier chapter I told you that the generic form of a parabolic equation was $y = x^2$. Actually, the generic form is clearly a little more complicated than that. Probably the most complete version, which is called *vertex form*, would look like this:

$y = a(x - h)^2 + k$

In this form, the constant **a** affects the "steepness" of the parabola and whether it opens upward or downward, **h** represents the horizontal shift of the vertex, and **k** represents the vertical shift of the vertex. Notice that this equation is similar in some ways to the one at the end of the last lesson that allowed you to graph circles with centers not at the origin.

Also notice that if **h** is equal to zero, the vertex form ends up looking like this:

$y = ax^2 + k$

... which is the generic form of the parabolas you worked with in Problems 11 through 17. If **k** is also zero, you've got:

$y = ax^2$

... which is the generic form of the parabolas you worked with in Problems 3 through 9. And finally, if **a** is equal to **1**, you have:

$y = x^2$

... the classic parabola.

31. Write a *Note to Self* about *parabolas and their graphs.* In a sense, this Note can be a combination of your answers to Problems 10, 14, and 20, or a description of the vertex form of a parabola and how it works. It should definitely include sketches of graphs!

Before I go on to talk about the geometric definition of a parabola, I'd like to say just a little more about parabolas from an algebraic perspective. As you'll discover in the next textbook in this series, the vertex form of a parabolic equation like this one:

$y = (x - 2)^2 + 3$

... can be rewritten like this:

$y = x^2 - 4x + 7$

Don't worry about how I did that yet, but do realize that the graph of the second equation is exactly the same as the graph of the first — it's still a parabola. Any time you see a two-variable equation where the highest power of the dependent variable is **2** and you aren't dividing anything by the dependent variable, the graph will be parabolic.

So all of these equations have parabolic graphs:

$$y = -7x^2 + 5x - 3$$

$$y = \frac{x^2 - 5}{2000} - x$$

$$y = (x - 3)^2 + 85x - 9$$

The work that you have done so far has focused on parabolas as we work with them and think about them in algebra. It's important to know that they also have a geometric definition. Remember from the last lesson that a circle was defined as "the set of all points in a plane that are the same distance from a fixed point in the plane" and you figured out the definition of an ellipse yourself. (I hope you came up with something like "the set of all points in a plane such that the sum of the two distances from each point to two fixed foci is the same.") Those were the geometric definitions of a circle and an ellipse. The geometric definition of a parabola is "the set of all points such that the distance from each point to a fixed focus is the same as its distance from a fixed line." That's probably a little hard to understand. Here's an illustration:

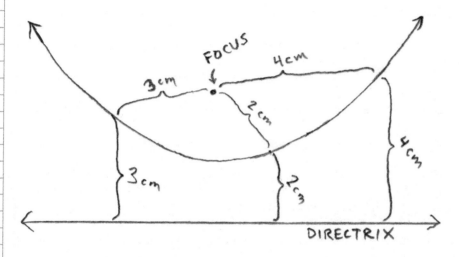

Notice that, no matter what spot you pick on the parabola, that point is the same distance away from the parabola's focus (which, like the foci of ellipses, is not actually *on* the parabola) and from the straight line (which is called the parabola's *directrix*).

There are several very practical (and clever) applications of parabolas that have to do with their foci. For example, a car's headlight (or a flashlight) is a parabolic mirror. If we were to slice a headlight in half, it would look something like this:

The actual light bulb is placed at the focus of the parabola. The effect of this is that the light rays leaving the bulb bounce off of the parabolic mirror and come out as a straight beam, like this:

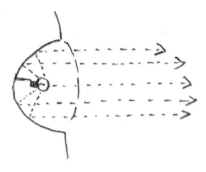

(You can easily take apart some kinds of flashlights in order to see their parabolic mirrors. Please don't try this on your parents' car.)

Satellite dishes use the same principal in reverse. Take a look at the next satellite dish that you see and notice that there is a gadget suspended above the dish, where I've drawn the arrow:

That gadget is the receiver for the dish. The dish itself is parabola shaped (the correct term for a three-dimensional parabola is a *paraboloid*), and the receiver sits — you guessed it — right at the focus of the paraboloid. So the dish has the effect of focusing the incoming rays right on the receiver. In cross-section it might look like this:

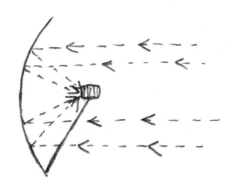

32. One last exploration having to do with parabolas. You've sketched parabolas using a series of points. It's also possible to approximate a parabola using a series of lines. Copy the figure below into your notebook, then connect each point to the point with the same number using a straight line. It will probably work best if you let each unit be two graph paper squares. Also, for best results, use a ruler or other straightedge to make your lines.

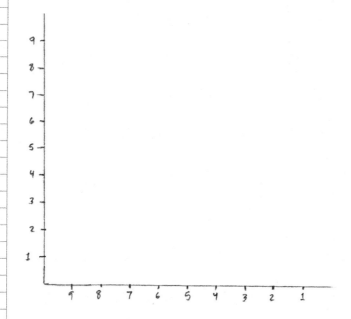

Notice that the curve that seems to appear looks like a parabola. It's possible to make all sorts of quite beautiful shapes using similar methods.

For example, copy and try this one on a larger sheet of graph paper, connecting only the points on adjacent axes:

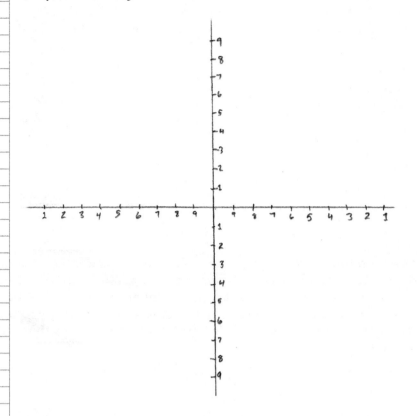

There are a million different variations on this. You can, for example, make straight lines between more legs, like this:

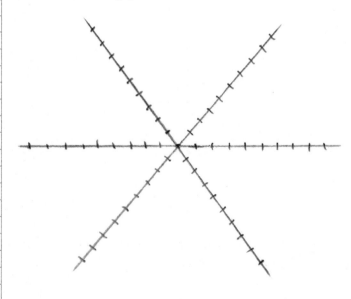

Chapter 4, Lesson 4

Or change the orientation of the lines, like this:

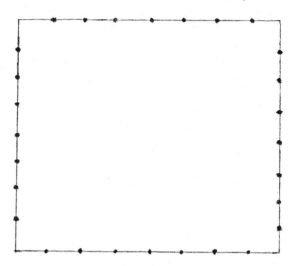

I got these ideas from a book called *Line Designs* by Dale Seymour, Linda Silvey, and Joyce Snider, which has tons of beautiful patterns in it. Unfortunately, as far as I can tell, that book is out of print. However, Dale Seymour has a more recent book (*Introduction to Line Designs*), as does Linda Silvey (*Designs from Mathematical Patterns*), so if you like these designs you might want to check those books out.

REVIEW

1. The number of hours left in the day is one-third of the number of hours that have already passed. How many hours are left in the day? (Solve this algebraically by letting x be the number of hours left in the day.)

2. Lighting experts recommend 150 to 200 watts of illumination for every 50 square feet of floor space. What is the minimum number of watts recommended for a room with a rectangular floor measuring 30 feet by 40 feet?

3. There are 1,760 yards to a mile and about 2.5 cm to an inch. To the nearest whole number, how many centimeters are in a mile? (You can use a calculator for this problem as long as you show your work clearly — including units!)

4. A pole 512 cm long is painted red, white, and blue in the ratio 3 : 4 : 1. What length of the pole is painted white?

5. Place >, <, or = signs between each pair to make the statement correct.

$$\frac{1}{6} \quad \frac{1}{7} \qquad \frac{15}{22} \quad \frac{15}{21} \qquad \frac{2}{3} \quad \frac{5}{9}$$

6. Linus donated $5 to the Save the Sea Monkeys campaign for every $4 donated by Greg. If they donated $1,800 altogether, how much money did Linus donate?

7. Audrey is at the Eiffel Tower buying a "J'aime Paris" T-shirt. The price is 20 euros, the sales tax is 19%, and the exchange rate is $1.50 to one euro. How many U.S. dollars will Audrey pay for her shirt?

Find the greatest common factors and least common multiples of the following sets of numbers:

8. $60n^3$
$350n^2$

9. $693m^3n^2$
$147m$

10. Sarah's Glove Hut has expanded to selling scarves — really expensive scarves made from luxury duck feathers. (Don't worry, no ducks are harmed in the process. They quite enjoy the grooming. And don't ask how you make weavable yarn out of duck feathers. It's a top-secret, patented process.) Sarah has found that she sells an average of 500 scarves per month at $100 per scarf. She has also realized that for every $5 she reduces the price she sells an extra 50 scarves each month. As you continue to lower the price, see what happens to her profit. Make a chart that shows the total number of scarves sold and total money Sarah will make if she sells them for $100, $95, $90, $85, $80, $75, $70, $65, or $60. (You can use a calculator to fill in the chart.) Sketch the shape of the function that shows how her income depends on the price of the scarves. What price should Sarah sell the scarves for to maximize her income?

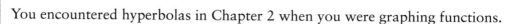

5 HYPERBOLAS

You encountered hyperbolas in Chapter 2 when you were graphing functions.

They are the ones that go with inverse functions, such as $y = \frac{3}{x}$ or $y = \frac{10}{x}$ or $y = \frac{500}{x}$.

The generic formula (or at least its basic version) is $y = \frac{a}{x}$. Hyperbolas look like this:

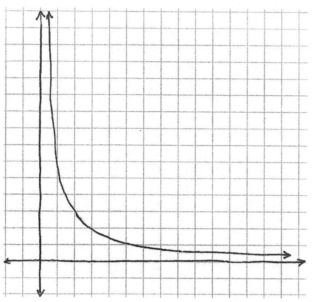

1. **As a little review, when you're dealing with an inverse function (and therefore a hyperbola), what happens to y as x gets very, very big? What happens to y as x gets very, very small (but stays positive)? When is y equal to zero?**

As I mentioned in Chapter 2, the graph above is only half of a hyperbola — it's called a *branch* of the hyperbola. The reason you only worked with one branch of a hyperbola is that you were using it to represent a real-world situation: the relationship between the volume of a balloon and the density of the air inside. There is no such thing as negative volume or negative density (at least not that I'm aware of!), so you only needed to deal with the positive branch of a hyperbola.

However, if you consider an inverse equation such as $y = \frac{1}{x}$ from a purely mathematical

standpoint (without worrying about what the two variables might represent), it's certainly

possible to have negative values for **x** and **y**.

2. In order to graph the other branch of a hyperbola, think about the simple inverse function $y = \frac{1}{x}$.

Make a coordinate graph. You're only going to graph points in the third quadrant — where both x and y are negative — so only make that part, and let four graph paper squares equal one unit on both axes.

Now copy and fill out the following chart for $y = \frac{1}{x}$:

x	-4	-3	-2	-1	-1/2	-1/3	-1/4
y							

(You'll need to divide by several fractions, but you can do that.)

Now put those seven points on your graph (two of them won't quite be at intersections of the graph paper lines) and connect them with a smooth curve, assuming that the curve will follow the same arc in both directions.

3. When you're dealing with an inverse function like $y = \frac{1}{x}$, what is the one value that x can never, *ever* have? What does this mean about the graph of an inverse function and the y-axis?

As you were probably able to figure out, the second branch of a hyperbola is a mirror image of the first.

Here's the graph of $y = \frac{4}{x}$:

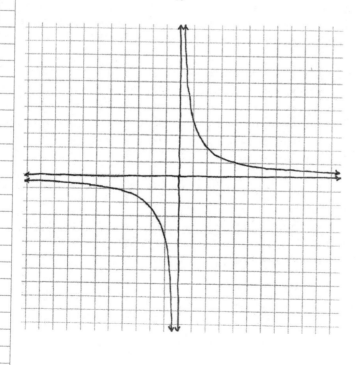

4. From the following graph, determine the equation of the hyperbola:

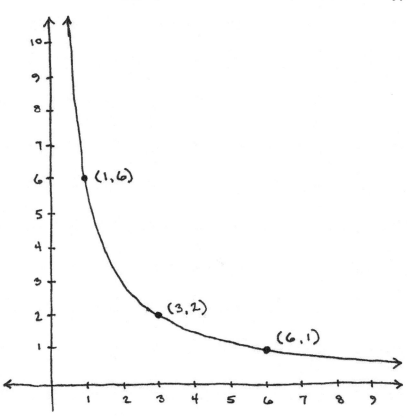

Let's take a look at the function $y = -\dfrac{1}{x}$.

5. Copy and fill in the following table for the function $y = -\dfrac{1}{x}$:

x	-4	-2	-1	-1/2	-1/4	1/4	1/2	1	2	4
y										

Now make a coordinate graph with a scale of 1 unit to 2 graph paper squares for both axes and graph those points on it. Connect them with smooth curves.

Chapter 4, Lesson 5

6. Just as you did in Problem 4, determine the equation of the hyperbola:

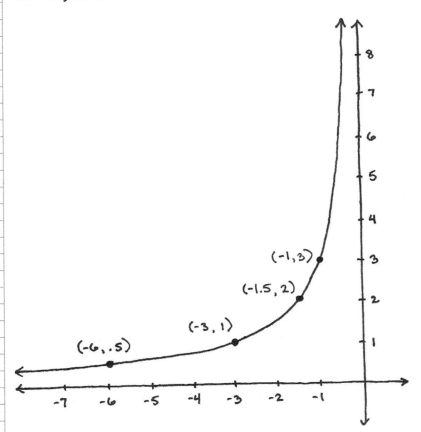

7. As you know, an inverse function of the form $y = \frac{a}{x}$ can never, *ever* cross the y-axis.

But, with a slight variation, it *can* cross the x-axis.

Consider the function $y = \frac{1}{x} + 1$.

Fill in the following table for that function:

x	-2	-1	-1/2	-1/4	1/4	1/2	1	2
y								

Graph those points on a set of axes with a scale of 1 unit : 2 graph paper squares. Keeping in mind what you know about hyperbolas, connect them with two smooth curves, keeping in mind that x still can't be equal to zero. (Right?!?!)

Make *sketches* of the following functions — get the overall shape right, but don't worry about individual points:

8. $y = \frac{-1}{x} + 3$

9. $y = \frac{1}{x} - 3$

Not surprisingly, it is also possible to move the graph of an inverse function to the left or the right so that it crosses the **y-axis**.

Consider the function $y = \frac{1}{x - 3}$.

10. For the function $y = \frac{1}{x - 3}$, what is the one value that x is never, *ever* allowed to have?

Use what you realized in Problem 10 to sketch the following functions:

11. $y = \frac{2}{x - 5}$
      12. $y = \frac{4}{x + 3}$

And it is possible to combine all of the things that you've just learned about hyperbolas...

Sketch the graphs of the following functions:

13. $y = \frac{1}{x + 2} + 3$
      14. $y = \frac{5}{x + 2} - 4$

15. $y = \frac{-3}{x + 4} + 1$
      16. $y = \frac{-2}{x - 5} - 3$

As you can see, the generic version of an inverse function really ought to look like this:

$y = \frac{a}{x + b} + c$

... rather than just this:

$y = \frac{a}{x}$

There is a second kind of equation that also gives you a hyperbola. It's a lot like the equations that give you circles or ellipses. Here's an equation for a hyperbola:

$\frac{y^2}{9} - \frac{x^2}{16} = 1$

17. What makes this equation different from an equation for an ellipse?

The graph of the equation $\frac{y^2}{9} - \frac{x^2}{16} = 1$ looks like this:

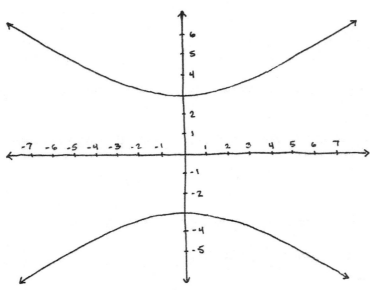

Although that graph is not tilted the same way as the graph of an inverse function,

it's still a hyperbola. I didn't ask you to try to graph $\frac{y^2}{9} - \frac{x^2}{16} = 1$ on your own because it's

really quite hard, but there's a way to make it a little easier.

Here is that same graph with a couple of features added. (They are not part of the graph of the equation, only visual aids to make it easier to graph.)

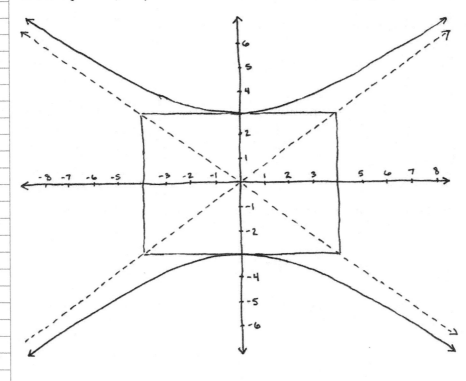

Chapter 4, Lesson 5

The dashed lines in that last graph play the same role that the **x-axis** and **y-axis** play in an inverse function: the branches of the hyperbola will get closer and closer to those two lines, but never touch them. They're called the *asymptotes* of the hyperbola. (I think that the asymptote is a really beautiful and rather poetic idea: the hyperbola gets infinitely close to it but never touches it.)

18. Now, in that graph the asymptotes are the diagonals of a rectangle.

Compare the dimensions of that rectangle to the equation $\frac{y^2}{9} - \frac{x^2}{16} = 1$.

What appears to be the relationship between the dimensions of the rectangle and the numbers in that equation? (It might help to look back at the work you did on ellipses.)

19. Based on your answer to Problem 18, draw a sketch (using a rectangle) of what you think the graph of $\frac{y^2}{16} - \frac{x^2}{9} = 1$ looks like.

Sketch the graphs of the following equations:

20. $\frac{y^2}{4} - \frac{x^2}{4} = 1$

21. $\frac{y^2}{1} - \frac{x^2}{9} = 1$

Look ahead to Problems 22 and 23, but don't do them yet. They have hyperbola-shaped graphs like the ones you just sketched, but with one important difference. To see what that difference is, look back at the hyperbola you just sketched:

$\frac{y^2}{1} - \frac{x^2}{9} = 1$

You can tell by looking at that equation that its graph will never touch the **x-axis**. If it did, **y** would be equal to zero. If you replace **y** with zero, the equation would look like this:

$-\frac{x^2}{9} = 1$

But no value of **x** could ever make that true, because **x²** is always positive, so $-\frac{x^2}{9}$ is always negative, and a negative number can't be equal to **1**.

Apply that logic to sketch graphs for Problems 22 and 23:

22. $\frac{x^2}{4} - \frac{y^2}{4} = 1$

23. $\frac{x^2}{9} - \frac{y^2}{4} = 1$

In the cases of circles, ellipses, and parabolas, there were algebraic definitions (that is to say, equations) for the three shapes, and there were also geometric definitions. I bet you won't be surprised to learn there's a geometric definition for a hyperbola.

A hyperbola, like an ellipse, has two foci (marked as two dots in this illustration):

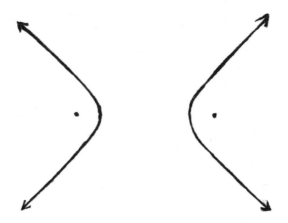

24. Just as I did for the ellipse, I'm going to ask you to figure out for yourself what the geometric definition of a hyperbola is. In order to do so, study the next illustration. This time I've marked three points that are all on the hyperbola (points A, B, and C) and I've marked the distances between those three points and the two foci. Here's your hint: The geometric definition of a hyperbola is quite close to the geometric definition of an ellipse — with one key difference! (By the way, I know three points make very little basis for inductive reasoning. Feel free to choose other points on the hyperbola and measure the distances between those points and the foci.)

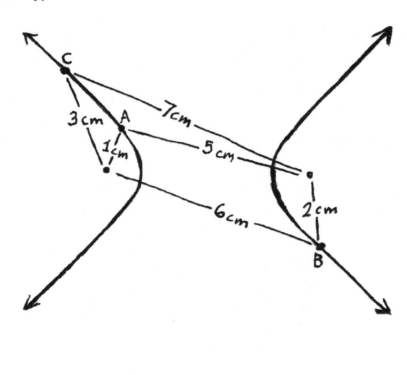

Chapter 4, Lesson 5

25. Now's a good time to make a **Note to Self** about *hyperbolas and their graphs.* Remember that there are two basic types of equations that you need to write about. Definitely include illustrations.

The title of this chapter is "The Conic Sections." Now it's finally time to find out why. Ancient Greek mathematicians knew about the four curves that you've looked at in the last few lessons: the circle, the ellipse, the parabola, and the hyperbola. Most especially, these curves were studied carefully by a man named Apollonius more than two thousand years ago. He called them *the conic sections.* In order to see why, I want you to imagine two cones with their tips touching. (If you like, you can imagine two ice-cream cones touching to make a sort of hourglass shape.) They would look something like this:

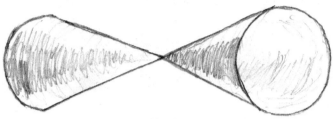

(Really, both cones ought to stretch out infinitely from the point where they meet. But I can't draw that.)

Now, imagine making a slice through these cones. (Or, if you prefer more mathematical language, imagine *passing a plane* through these cones.)

One way to make such a slice would be to come straight down like this:

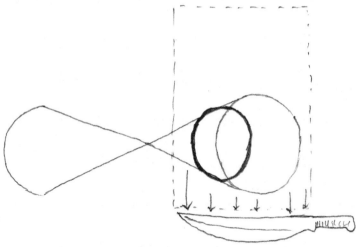

26. Imagine the shape that is made at the edge where the cone has been sliced off. What shape would it be?

A second way to slice would be still to cut right through one of the cones, but to do so at an angle, like this:

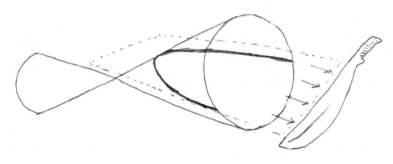

27. **What shape would the edge of the sliced cone be in this case?**

You could make a slice like that, all the way through one cone, at a variety of angles. If you kept making the angle of the slice less and less steep, you'd eventually reach a point where, instead of cutting through to the other side of the cone, you'd be cutting *along* the cone, like this:

(Remember, these cones are really supposed to be infinitely long, so the cut edge should be infinitely long, too.)

28. **This time what shape does the sliced edge make?**

Again, you could keep flattening the plane of your slice more and more until, at some point, you'd be cutting through both of the cones and you'd have two opposing sliced edges, like this:

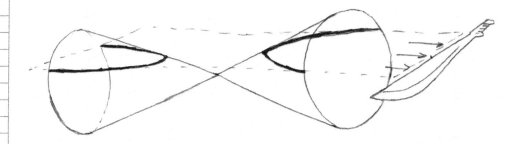

29. **This time what shape is made by the sliced edges?**

By cutting these cones, it's possible to get many different varieties of each curve, but there are only four different kinds of cuts you can make. Each type of cut produces one of the four different curves you've been working with in this chapter.

30. **I lied to you just a little bit in that last paragraph. There are actually two ways to cut the cones that don't make conic sections. Can you figure out what they are?**

There are some pretty good pictures and animations on the internet of the conic sections and how they work. Here are two that you might want to check out:

http://rowdy.mscd.edu/~talmanl/HTML/ConicSections.html
http://id.mind.net/~zona/mmts/miscellaneousMath/conicSections/conicSections.htm

The four conic sections are two dimensional, but what happens when you add a third dimension? A circle becomes a sphere. An ellipse becomes a kind of egg shape. The others are a little harder to picture in your head. But it turns out we can use origami to model one of the possible forms — the *hyperbolic paraboloid*, or *hypar*. This beautiful infinite surface was discovered mathematically in the 17th century, but most people were probably already familiar with the shape: it looks like a saddle. A Pringles crisp takes this shape, too. Architects have used the form since the 1950s to create dramatic, sculptural roofs.

31. **The pleated hypar folding you're about to make was originally designed by a fellow named John Emmet in England. You'll need some large origami paper, at least 8.5 inches square. You can use regular computer or notebook paper if you first cut off one end to make a square, but origami paper is thinner and will help you make the many precise folds more accurately. Follow the instructions on the next page, which are modeled on those created by Erik and Martin Demaine and Anna Lubiw, a team of computer scientists at the University of Waterloo in Canada. (You might be surprised to learn that scientists would use art to help people visualize hypars and even more complex forms. If you study chaos theory at some point you will learn about a fractal structure called the Lorenz attractor that illustrates complicated effects of convection in the atmosphere. Some mathematicians recently discovered that it was possible to crochet an accurate model of the Lorenz attractor and they published instructions in a journal called The Mathematical Intelligencer.)**

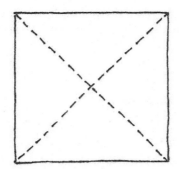

Crease the diagonals ...

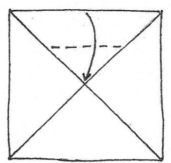

fold the top edge to the center point, creasing only between the diagonals ...

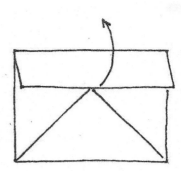

unfold ...

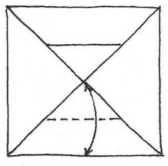

repeat on the bottom (fold/unfold) ...

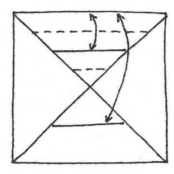

fold and unfold on 1/4 and 3/4 marks ...

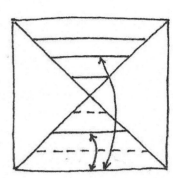

repeat on the bottom ...

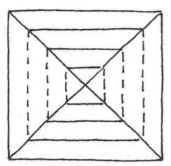

repeat on left and right sides ...

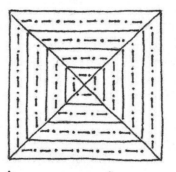

turn over and crease between the squares in the opposite direction.

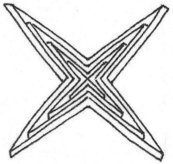

Folding all the creases forms an X shape.

Partially opening it forms a hypar !

–––– valley fold
–·–· mountain fold

REVIEW

1. If Leah is 6 years older than Sue and John is 5 years older than Leah and the total of their ages is 41, how old is Sue?

2. To make lemonade for the weasels, armadillos, and cyclopes I've invited to Greg's surprise birthday party, we need the following ingredients:
   6 cups of lemon juice
   14 cups of water
   5 cups of sugar

   a. What is the ratio of lemon juice to water to sugar?
   b. Use an illustration to represent the ratio.
   c. What fraction of our lemonade is sugar?
   d. Sugar makes up what percent of our lemonade?

3. Byron and Erin have the same number of antique apple-corers. If Byron gives 1/3 of his apple-corers to Erin, what will be the ratio of Byron's apple-corers to Erin's? (Hint: Drawing a picture is a very effective way of solving this problem quickly!)

4. Abby needs to make a cake and some cookies. The cake requires 3/8 of a cup of sugar and the cookies require 3/5 of a cup of sugar. She has 15/16 of a cup left in her bag. Does she have enough sugar? If not, how much more does she need?

5. Abe, Barry, and Carlos have 256 model moose altogether. (I leave it to you to determine whether these are models of moose or exceptionally moose-ish actual moose.) The ratio of Abe's moose to Barry's moose is 4 : 3. Barry has 14 more moose than Carlos. How many moose does Abe have?

6. The original price of a stylish raincoat was $50. The store deducted 20%, and then deducted an additional 20% from the reduced price. How many dollars more would Fred have saved if the store had simply reduced the original price by 40%?

7. Simplify the following expressions. You'll be far happier if you start by prime factorizing the numbers involved.

   $$\frac{(24)(5)(13)}{(25)(39)(12)} \qquad \frac{(19)(4)(4)}{(16)(38)(2)} \qquad \frac{(17)(18)(2)}{(9)(34)(2)}$$

8. In a group of 1,000 people, 40% believe corn flakes are a superior breakfast cereal, and 30% of the corn-flake lovers also favor hot dogs over bratwurst. Of those corn-flake aficionados who prefer hot dogs, 15% like to dress as pirates. How many people in this group love corn flakes, hot dogs, and peg legs?

Write equations to go with the following tables and tell what the overall shape of each graph would be (straight line, parabola, or hyperbola). These may be a little more challenging than the ones you've done before.

9.

x	0	1	2	3	4	5
y	7	4	3	4	7	12

10.

x	3	5	7	9	11
y	1	2	3	4	5

11.

x	8	10	13	18	28	58
y	6	5	4	3	2	1

12. There's a certain kind of lion that is perfectly logical, always very hungry, and has a peculiar disorder: every time it eats, it falls fast asleep. This kind of lion is also cannibalistic and cowardly: it's just as happy to eat another lion as it is to eat anything else — but it will only eat another lion if that lion is asleep. The question: If there are 57 of these peculiar lions together in a cave and a sheep wanders into the cave, will the sheep get eaten? Why or why not?

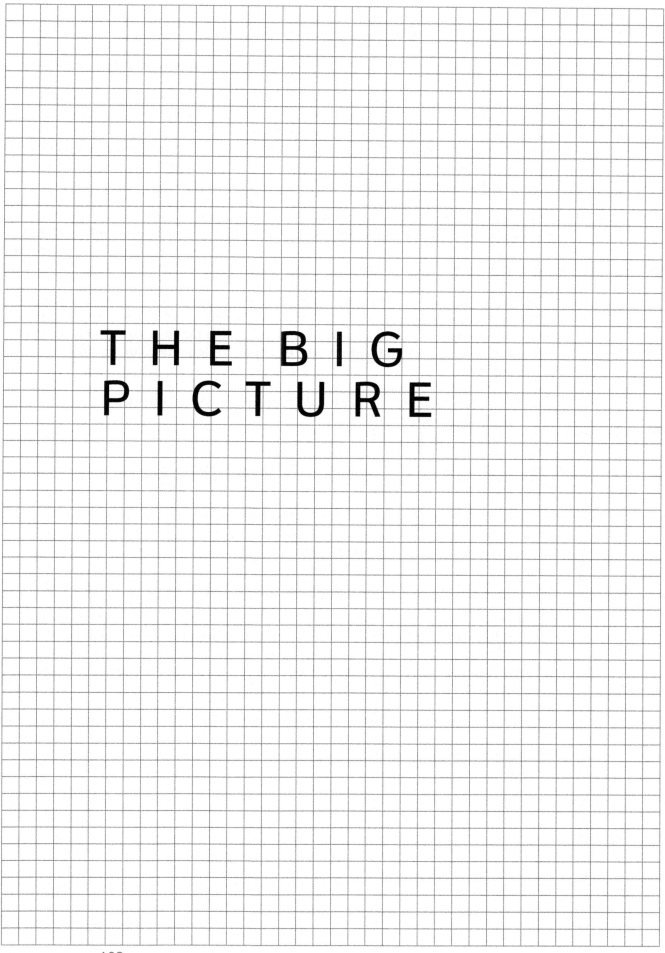

THE BIG PICTURE

You've done a lot of work in this textbook and (I hope) learned a great deal. Before you move on, I'd like to ask you to take a step back and consider the big ideas from this book. I won't give you a ton of problems to do in this epilogue, but please read it carefully and ask questions if you don't understand it. I think it's very important.

You know from *Jousting Armadillos* that variables can be used to solve problems. In a sense, the simultaneous equations work from this book is an extension of that kind of work: you're faced with a problem in which there are one or more unknown numbers and you use variables to figure out what those numbers are. Here's one last ratio problem that can be solved in that manner:

1. **In a jar filled with beads, 2/5 of the beads are blue, 1/3 of them are red, and the rest are green and yellow. The total number of red, green, and yellow beads is 126. There are 3/4 as many green beads as yellow beads. How many beads of each color are there?**

(I think this problem is hard to solve using the picture ratio method. I solved it by using five variables: b, r, g, y, and x for the total. Then I wrote down everything I knew as equations using those variables and used simultaneous equation techniques to figure out what the variables were. For example, the equation for the total is b + r + g + y = x.)

Great. In that problem there is a specific, single set of answers that works. Algebra can be a powerful problem-solving tool.

But I've hoped to show you in this book that there is more to algebra than figuring out how many beads there are in a jar or how old so-and-so will be in six years or how fast so-and-so is running.

The heart of this textbook, as far as I'm concerned, is the two-variable equation. A two-variable equation (considered on its own and not as part of a set of simultaneous equations) is a completely different sort of beast.

A two-variable equation, such as $y = 3x - 2$ or $2x - y = 6$ or $y = \dfrac{6}{x - 3}$ or $x^2 + y = 6$ or the

many, many others you've encountered, is very different from a problem-solving situation such as Problem 1 where you're looking for a specific solution. You've learned that a two-variable equation has an infinite set of solutions that can be displayed on a two-dimensional graph and that it can be used to represent changing relationships in the real world. (Remember those candles?)

You are now familiar with a number of different "families" of two-variable equations. I'd like to go over those families one at a time. In each case, I'll give an equation with its graph and say a few words about it.

$$\frac{x^2}{9} + \frac{y^2}{9} = 1$$

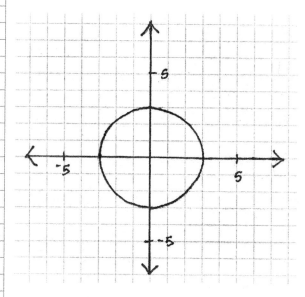

This belongs to the family of circles. In this case, it's a circle with its center at the origin and a radius of **3**. Remember that you have worked with circles that have their centers at spots other than the origin. They have equations such as **(x - 4)² + (y - 2)² = 4.**

$$\frac{x^2}{9} + \frac{y^2}{4} = 1$$

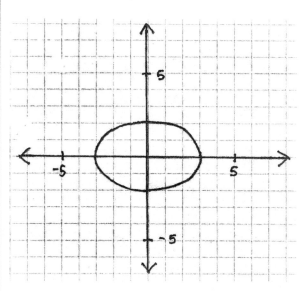

This belongs to the family of ellipses. In this case, it's an ellipse with a major axis of **6** and a minor axis of **4.**

$$\frac{x^2}{4} - \frac{y^2}{9} = 1$$

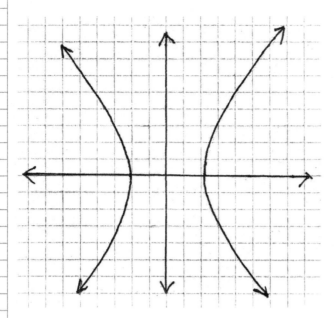

I'm going to call this family "non-functional hyperbolas." As far as I know, that's not a real mathematical term, but I'm calling them that because, well, they're not functions and it's important to distinguish between them and the other family of hyperbolas that *are* functions.

If you have an equation like $\frac{x^2}{9} - \frac{y^2}{4} = 1$, the branches of the hyperbola open up and down instead of to the left and right.

The three families I've mentioned so far — circles, ellipses, and non-functional hyperbolas — have two important qualities in common. First, none of them are functions. All of their graphs would fail the vertical line test that you learned in Chapter 2. Second, in all of their equations both variables are squared. This is true even in a case like $(x - 4)^2 + (y + 2)^2 = 4$; other stuff is being squared *with* the variables, but both variables are definitely being squared.

If I asked you to graph any equation that belonged to the circle, ellipse, or non-functional hyperbola families, you could do it. Even if you forgot all of the tricks that you learned in Chapter 4, you could still do it: you would just make a function table, find **x** and **y** pairs that made the equation true, and you could figure out the graph. Instead, I'm going to ask you to take a step back and realize this: *any time you have a two-variable equation where both variables are squared, its graph will belong to one of these three families.* (As long as you're not dividing by either of the variables — we'll set that case aside.) When you see an equation like $(x - 6)^2 + 85 - (y + 2)^2 = 603$, I don't expect you to be able to picture its graph in your head. (I can't!) But I would like you to be able to say to yourself, "Ah. Both variables are being squared. This thing is definitely not a function. It's a circle, an ellipse, or a non-functional hyperbola."

You should be able to say the same thing about any of these equations:

$$x^2 = y^2 - 65$$
$$(x - 3)^2 = 24 - y^2$$
$$0 = (y - 2)^2 + x^2$$

... or any others like them.

Okay, let's move on to the families of functions that you know. I think it's probably fair to say that functions are more important than non-functional two-variable equations, at least from an algebraic point of view. That's because functions allow you to make reasonable predictions. Remember, you could tell an intelligible story to go with this graph:

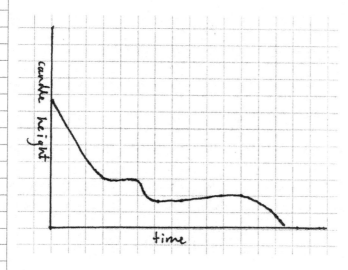

... but this graph was nonsensical:

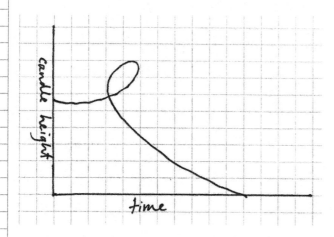

... because the candle would have to be three different heights at the same time.

The Big Picture

So here are the families of functions that you know.

y = 2x + 3

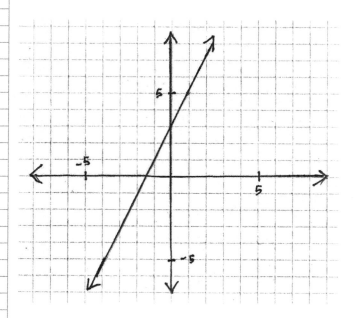

This, of course, belongs to the linear family. As you know very well, if the equation is in **y-equals** form and the **x** part is *negative*, the line slopes *down* from left to right.

Again, I won't ask you to graph any linear functions right now. But realize this: *any time you have a two-variable equation where neither variable is squared and you're not dividing by either variable, the graph is a straight line.* This is true because you can always change that equation into **y = mx + b** form. So, when you see any of the following equations, you should say to yourself, "Ah-hah. A straight line."

y = -6x + .75

x + y = 2

(y - 3) + x = 6x - 2

$-x = \dfrac{2y}{3}$

x + x + x + x + y = 10

$\dfrac{x}{1008} - 5 = \dfrac{y}{1009}$

Here's the next family:

$$y = x^2 + 3$$

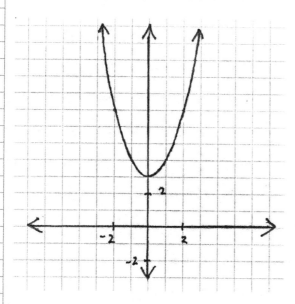

This belongs to the parabolic family. Remember that if the equation is in **y-equals** form and the x^2 part is negative, the parabola opens downward instead of upward. Also remember that you can move the vertex of the parabola around when you have an equation like $y = (x - 2)^2 + 5$.

Once again, I won't ask you to graph any parabolas, but I will ask you to realize an important fact: *any time you have a two-variable equation where the independent variable is squared and the dependent variable isn't and you're not dividing by the independent variable, the graph is a parabola.* (Actually this is only true as long as there isn't a higher power of the independent variable, such as x^3 or x^7. But you'll learn about that in the next textbook, so don't worry about it yet.) When you see any of the following equations, you should say to yourself, "Gee willikers, Batman (because I know you address yourself as Batman), a parabola!"

$$y = x^2 - 5$$
$$y = -10(x - 3)^2 + 5$$
$$y = 5x^2 + 3x - 9$$
$$\frac{y}{53} = \frac{x^2 + 9}{2} - .0004$$
$$1001y - 2y + 8x^2 = 6x$$
$$\frac{y}{.8} = x^2 + 3x^2 + 2x^2 + 5000x$$

And now for the last family of hyperbolas (or, if you prefer, "functional hyperbolas," in order to distinguish it from the other family).

An example would be $y = \frac{6}{x}$ and its graph would look like this:

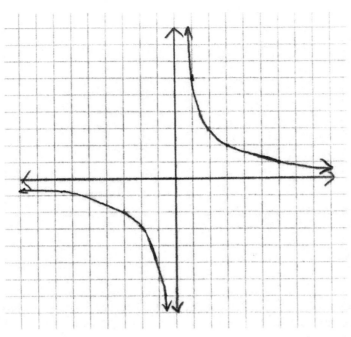

As you know, if the equation is in **y-equals** form and the $\frac{a}{x}$ part is negative, the hyperbola

is flipped so that its branches are in Quadrants II and IV instead of Quadrants I and III.

And you can shift the hyperbola around by doing things like this: $y = \frac{1}{x-2} + 6$.

Once again, the thing that I want you to realize is this: *any time you have a two-variable equation in which you are dividing something by the independent variable, the graph will be a hyperbola.* (We'll set aside equations where you're dividing by the square of the independent variable for right now.) So, when you see any of the following equations, you should declare to yourself, "Arrr, shiver me timbers, 'tis a hyperbola!"

$y = \frac{6}{x} + 19.2$

$y = \frac{10}{x + x + x}$

$y = \frac{.005}{6x}$

$y = \frac{-3}{2x} + 6x - 5$

$52y - \frac{15}{3x + 2} = 0$

$\frac{17}{x - 3,000} = \frac{y}{5} + 4y + \frac{2}{3}$

As I warned you, that was a lot of reading. But I hope it was clear to you because, as I said, I think it's quite important. So, one last problem set and you'll be ready to move on to the next book....

For each of the problems below, say whether the graph would be a straight line, a parabola, a functional hyperbola, or a non-functional curve. (You don't need to distinguish between circles, ellipses, and non-functional parabolas.) Assume that x is the independent variable, y is the dependent variable, and any other letters are constants.

2. $y = 7x^2 - x + 2$

3. $\frac{y}{10} + x = 6$

4. $3y + y - 2 = (x + 5)^2$

5. $4x^2 - 2 = 5 - \frac{y^2}{5}$

6. $x + 3 + x + 2 + y = 0$

7. $\frac{10}{x} - 3y = .06$

8. $15 + 8y = \frac{3}{x - 2}$

9. $\frac{x^2}{6} - y^2 + 6 = 5x^2$

10. $\frac{2x}{3} - \frac{5y}{3} = 8$

11. $5(x - 3)^2 = y + y + y$

12. $.6x = 2,005y$

13. $(2 + 3x)^2 - 6 = (y - 6)^2$

14. $y + 5y = \frac{13}{10 + x}$

15. $y = \frac{m}{5} + 3x$ (Remember, you can assume letters other than x and y are constants.)

16. $y = (m - x)^2 + c$

17. $y = \frac{a}{3x + b} - c^3$

18. $y + mx^2 - x = n$

19. $mx^2 + ny^2 = q$

20. $y - \frac{a + b + c}{x} = d$

21. One last **Note to Self** for this book. You can call it "The Big Picture" or "Recognizing Functions" or "What I Learned in *Crocodiles & Coconuts.*" Call it "Dr. Xavier Moose-spackle's Fabulous Note-to-Self Extravaganza" if you like. What it needs to include is the information necessary to do Problems 2 through 20. *How can you recognize what kind of function you're dealing with when you look at an equation?*

Excellent. You've come a long way toward mastering introductory algebra (Algebra I) and you deserve to be proud of the work you've done. In the final volume of this series, you'll continue to build on the skills you've developed so far as you explore polynomials and quadratics.

CPSIA information can be obtained
at www.ICGtesting.com
Printed in the USA
LVOW02s2156070817
544127LV00001B/88/P